Christel Anne Ross

Invasion Success by Plant Breeding

W0037545

VIEWEG+TEUBNER RESEARCH

Christel Anne Ross

Invasion Success by Plant Breeding

Evolutionary Changes as a Critical Factor
for the Invasion of the Ornamental Plant
Mahonia aquifolium

With forewords by Prof. Dr. Helge Bruelheide
and Dr. Harald Auge

VIEWEG+TEUBNER RESEARCH

Bibliographic information published by the Deutsche Nationalbibliothek
The Deutsche Nationalbibliothek lists this publication in the Deutsche Nationalbibliografie;
detailed bibliographic data are available in the Internet at http://dnb.d-nb.de.

Dissertation der Martin-Luther-Universität Halle-Wittenberg, 2008, u.d.T.:
Evolutionary changes by plant breeding – a critical factor for the invasion success
of the ornamental plant *Mahonia aquifolium*

1st Edition 2009

All rights reserved
© Vieweg+Teubner | GWV Fachverlage GmbH, Wiesbaden 2009

Reader: Anita Wilke

Vieweg+Teubner is part of the specialist publishing group Springer Science+Business Media.
www.viewegteubner.de

Cover design: KünkelLopka Medienentwicklung, Heidelberg
Printing company: STRAUSS GMBH, Mörlenbach
Printed on acid-free paper
Printed in Germany

ISBN 978-3-8348-0792-2

Foreword

Christel has been intrigued by the phenomenon of invasions since her studies as an undergraduate student in botany at Goettingen University where she took several of my courses and where I supervised her diploma thesis. Her diploma thesis already addressed the possible impact of hybridization for the invasiveness of plant species. By using molecular markers, she studied North American and European *Rhododendron* species. We were also in close contact while she was working on her PhD thesis at the Department of Community Ecology at the Helmholtz-Centre for Environmental Research UFZ in Halle. Having been one of the reviewers of her PhD thesis, I readily agreed when she asked me to write a short preface to this publication.

While the main line of research on the role of evolutionary processes for plant invasions has mainly been on the response to a different natural selection pressures exerted by the abiotic and biotic site factors of the new environment, Christel has asked to which degree breeding efforts might have contributed to such pressures. She chose a very apt study object to address this topic, *Mahonia aquifolium*, a species native to North America and introduced to Europe as an ornamental plant, together with some other species of the same genus. Christel's basic question was whether invasive populations of *Mahonia aquifolium* in Europe originate from planted cultivars or from hybrids with *M. repens* und *M. pinnata*. She certainly deserves the credit to have studied this question with a broad methodological approach, using neutral genetic marker as well as common garden and transplant experiments, and with a remarkable personal devotion.

Although not all of her results confirmed her hypotheses, for which the reader has to refer to the nicely written articles and the summary in this booklet, Christel's work has delivered an important contribution to the field of invasion research. The reader will find answers to how the selection process influenced the genetic diversity of *Mahonia*, which traits of the species were affected and whether adaptation to local environmental conditions were involved. Certainly, not all questions have been given a final answer, which is to be expected even from such an excellent PhD thesis. The remaining open questions might be even more intriguing than the ones Christel started with in her career. However, this is what science is actually about, and I hope that readers of her work will be as intrigued as she has been and become inspired to carry on with this topic.

Prof. Dr. Helge Bruelheide

Foreword

By human activities, such as international trade, tourism, horticulture as well as fur, food or timber production, exotic plants and animals have intentionally or unintentionally been transported across major biogeographic barriers. In part, these exotic species have been able to establish self-sustaining populations and have spread into the new area. This biogeographic process, triggered by human activity, is called a biological invasion; it takes place at unprecedented spatial and temporal scales and is unique in the number of species dispersing. Biological invasions may generate large economic costs and are one of the major threats to biodiversity. An improved understanding of the processes behind invasions will, therefore, allow us to predict future invasions, to assess their impact, and to develop management strategies. On the other hand, invasions can also be considered as grand but unplanned experiments that can contribute to our basic understanding of biogeographical, ecological, and evolutionary processes.

One important question in the study of biological invasion is: What makes a species a successful invader? Only during the past decade, awareness has been rising that evolutionary processes may greatly contribute to the success of invasive species. Since then, more and more evidence has accumulated that the ability for evolutionary adjustments to novel environments may play a key role for successful invasions. Research on evolutionary changes in invasive plants has primarily investigated the role of natural selection. However, the majority of plant invaders have been imported intentionally, mostly as ornamentals for landscaping and gardening. Therefore, the effect of plant breeding, including hybridization and artificial selection, has to be taken into account when studying evolutionary changes in invasive plants.

Christel Ross' publication significantly contributes to our knowledge on the evolutionary importance of horticulture for the invasion success of exotic plants. Based on a case study on a widely-used ornamental shrub, *Mahonia aquifolium*, the author investigates the genetic relationship between native populations, planted cultivars, and invasive populations. Combining molecular analyses and a common garden experiment, Christel Ross demonstrates a strong genetic differentiation between native and invasive

Mahonia populations, which is likely to be caused by plant breeding. Using reciprocal transplant experiments, she, finally, investigates to what extent local adaptation to habitat conditions within the introduced range may have contributed to the species' invasion success. Considered together, the results presented here show that humans have obviously produced a successful invader by themselves – which is probably not only true for *Mahonia* but also for many other invasive species that were originally introduced as ornamentals. The present case study is thus not only important for our understanding of the role of evolution in plant invasions but is also of high practical relevance. Therefore, I am convinced that this publication will be of great benefit to all ecologists who are interested in the mechanisms behind biological invasions.

Dr. Harald Auge

Preface

This PhD thesis was written at the Martin-Luther-University Halle-Wittenberg and the Helmholtz-Centre for Environmental Research – UFZ in Halle. I am very grateful for having been given the opportunity to complete my thesis at these excellent institutions and to work together with such great colleagues. My special thanks go to my supervisors and co-authors of a number of articles Dr. Harald Auge and Dr. Walter Durka, for their everyday support, and to Prof. Dr. Helge Bruelheide for his guidance and encouragement.

This thesis is a cumulatively composed work. The contents of this thesis have been published in the following international journal articles.

Christel Roß & Walter Durka (2006) Isolation and characterization of microsatellite markers in the invasive shrub *Mahonia aquifolium* (Berberidaceae) and their applicability in related species. *Molecular Ecology Notes* 6, 948-950.

Christel A. Ross, Harald Auge & Walter Durka (2008) Genetic relationship among three native North-American *Mahonia* species, invasive *Mahonia* populations from Europe, and commercial cultivars. *Plant Systematics and Evolution* 275, 219-229.

Christel A. Ross & Harald Auge (2008) Invasive *Mahonia* populations outgrow their native relatives. *Plant Ecology* 199, 21-31.

Christel A. Ross, Daniela Faust & Harald Auge (2009) *Mahonia* invasions in different habitats: local adaptation or general-purpose genotypes? *Biological Invasions* DOI 10.1007/s10530-008-9261-y.

Christel Anne Ross

Contents

Mahonia invasions in different habitats: local adaptation or general-purpose genotypes?

Summary

Since invasive species are a major threat to global biodiversity, many countries, including Germany, are committed to reducing invasions by animals and plants. To successfully manage and control invasive species, it is essential to understand the causes of successful invasions as well as the underlying processes. A major proportion of successful invasive plants are cultivated plants. Yet only a few authors have investigated breeding and cultivation as a reason for successful invasions. Indeed, plant cultivation may affect invasions in different ways. First, suitable individuals of the native populations, which are particularly fit, may be introduced into the new area. Nevertheless, genetic variation is supposed to be reduced by selective introduction which results in a genetic bottleneck. Second, the majority of cultivated plants are planted in high quantities, and therefore propagule pressure to adjacent habitats and the probability of establishment in natural habitats are high. Third, cultivated plants are bred for advantageous traits, frequently using hybridisation with related species. Since interspecific hybridisation may lead to new genotypes and a high genetic variation, it can counteract harmful genetic bottlenecks. It has been repeatedly shown that rapid evolution plays an important role in plant invasions. The time lag between introduction and invasion of a plant species may be related to evolutionary changes that enable the introduced plant to establish self-sustaining populations under the new environmental conditions. That may either be due to local adaptation to the new environmental conditions or by evolving high phenotypic plasticity. Most studies that have investigated evolutionary changes in invasive plants have looked either for genetic bottlenecks, or for the Evolution of Increased Competitive Ability after release from specialist herbivores (EICA hypothesis). Because evolution in cultivated plants is fostered by breeders who artificially select genotypes that may be successful not only in gardens but also in native habitats, it is likely that breeding enhances the invasion success. *Mahonia aquifolium* is a well-suited ornamental plant to investigate this hypothesis. It is a shrub, native in North America and invasive in different natural habitats in Central Europe. During breeding, it was hybridised with the closely related species *M. repens* and *M. pinnata*. These two species are also native in North America.

Since many of these hybrids are planted in gardens, parks and along streets, and the seeds are dispersed by birds, the propagule pressure on adjacent habitats is high. It is supposed that invasive *Mahonia* populations originated from these cultivated populations and consist of hybrids between the three related species.

In this thesis, I investigate the following questions: 1) Are invasive *Mahonia* populations genetically different from native populations in neutral and quantitative traits? 2) Did a genetic bottleneck during the introduction reduce genetic variability or is genetic variability high due to hybridisation? and 3) Are invasive *Mahonia* populations locally adapted to their new environments? I explored native and invasive populations and cultivars using neutral genetic markers to find out if genetic changes occurred, and to analyse the genetic diversity in native and invasive populations. By means of a common garden experiment, I investigated the ecological relevance of genetic changes, and by different reciprocal transplant experiments, I tested for local adaptation of invasive populations.

My results show that invasive *Mahonia* populations have a high genetic diversity similar to the diversity in native populations. I found, in addition, that invasive populations were genetically different from native populations of all three species, but were most similar to *M. aquifolium*. Genetic analyses revealed that native *M. repens* populations are geographically isolated; *M. aquifolium* populations and *M. repens* populations, which were sampled in the northern part of their range, could not be separated by microsatellite markers. This weak differentiation between *M. aquifolium* and *M. repens* complicated the detection of hybrids. I found hybridisation of *M. aquifolium* and *M. pinnata* in cultivars, but not in invasive populations, and hardly any hybridisation of *M. aquifolium* and *M. repens*, either in cultivars or in invasive populations. The differences between invasive and native *Mahonia* populations were ecological relevant, as they could be found in quantitative traits. In the common garden experiment, invasive populations grew more vigorously in comparison to native *M. aquifolium* and *M. repens* populations. The amount of heritable variation, however, was similar in native and invasive populations. Furthermore, I could not detect genetic adaptation to different habitats using reciprocal transplant experiments.

Invasive *Mahonia* populations are genetically different from native populations, and this differentiation cannot be attributed to a genetic bottleneck or to a release from specialist herbivores. However, it was not possible to show unambiguously that the genetic differentiation was a result of hybridisation. Nevertheless, plant breeding and cultivation play an important role in *Mahonia* invasion at least by planting large quantities of individuals and a large range of cultivars. I furthermore suggest that breeding of especially fit cultivars plays a role in the invasion process, because I could show that not all cultivars are similarly aggressive invaders of natural habitats. The lack of genetic adaptation to different habitats and the finding that offspring of all populations grew best in the same habitat indicate a high phenotypic plasticity in invasive populations, at least for early life stages.

Chapter 1

Introduction

The invasion of new habitats by animals and plants represents a basic natural process as well as an anthropogenic phenomenon (Rejmanek 1996). However, the term "Biological Invasion" is used by most authors when species are spreading in a new area mediated by human activities (Auge et al. 2001; Richardson et al. 2000; Kowarik 1995; Pysek 1990). The distribution of species is, among other things, the result of natural dispersal that is limited by geographical or biological barriers. As human trade and transport increase, species can overcome these barriers and invade new areas. Thereby, the transport of species may either occur by accident, for instance in ballast water of ships or in packing material, or deliberately as introduction of cultivated plants or biocontrol species. The number of species that have been translocated has increased dramatically in the last 200 years (Mack et al. 2000; Williamson 1996; Drake et al. 1989). Out of 10 introduced species, one species may establish in the new area, and out of 10 established, one species likely causes problems in the new area ("tens rule", Williamson and Fitter 1996). Today, biological invasions are considered one of the greatest threats to global biodiversity (Millenium Ecosystem Assessment 2005; Dukes and Mooney 1999). Native species can be suppressed or replaced by invasive species either directly by competition, or indirectly by alterations of the ecosystem as well as by alteration of the relations between organisms (e.g. Sperry et al. 2006; Belnap et al. 2005; Callaway and Aschehoug 2000). The dimension of the threat by invasive foreign species is quite perceptible. For instance in the South African bushland, 80 % of endangered species have become rare as a consequence of invasive species (Armstrong 1995). In addition to ecological problems, invasive species cause considerable economic costs. For the USA alone, Pimental et al. (2000) estimated $137 billion per year for economic loss. Thus, forecasting future invaders will help to arrest those invasions and is essential for maintaining biodiversity as well as saving enormous economic costs. Indeed, forecasting future invasion will not be possible until fundamental research in the field of invasion biology has identified how species become invasive and which processes are important for invasion success.

1.1   The collaborative project INVASIONS

In Germany, invasive species cause ecological and economic problems (Reinhardt et al. 2003). The economic costs caused by 20 of the most notorious invasive species in Germany add up to an annual sum of about € 167 million, according to a recent study on behalf of the Federal Environmental Agency (Reinhardt et al. 2003). In the Convention on Biological Diversity (CBD), Germany committed itself to "prevent the introduction of, control or eradicate those alien species which threaten ecosystems, habitats or species" (article 8). Each signed nation is asked to develop a concept to deal with invasive species. The German Federal Ministry of Education and Research funded a project named INVASIONS – The invasion potential of non-native species – identification, evaluation, risk management (project-homepage http://www.ufz.de/ index.php?de=2773). The aim of the project was the identification of factors that facilitate invasions by plant species and the development of a scheme to estimate economic consequences. Further, future requirements for legal regulations concerning the differential treatment of non-native plant species according to their invasion potential and cost-effectiveness of measures should be suggested by the scientists. Three groups of biologists worked on (1) the identification of traits and introduction pathways of successful invasive plant species, (2) the importance of plant breeding for invasion success based on a case study, and (3) the diversity of native phytophagous insects on invasive plants and their effects on invasion success. Two further groups, formed by economists and legal experts, (4) developed criteria for the evaluation of non-native species to be used by decision makers, and (5) evaluated current legislation and developed improvements for future laws concerning invasive species. The study covered by this thesis is part of the INVASIONS project. It investigated the role of plant cultivation and breeding, which are socio-economic processes that contribute to invasiveness of plant species. The study was conducted using the invasive ornamental plant *Mahonia aquifolium* PURSH. (NUTT.) as a model system.

1.2   Evolutionary processes in plant invasion

Although many traits and processes that facilitate invasiveness have been investigated in the past, neither generally valid causes of biological invasions nor generally applicable predictors for future invasions have yet been identified (Mack 2000; Mack et al. 2000; Crawley 1987). In recent years, the evolutionary potential of introduced species has been found to play a key role in their invasion success (Bossdorf et al. 2005; Sakai et al. 2001; Caroll and Dingle 1996). Evolutionary changes have been detected in several invasive plant species (reviewed in Bossdorf et al. 2005). They may occur as a result of different processes associated with invasions. At first, genetic diversity may be decreased after a genetic bottleneck and genetic drift during introduction (Barrett and Richardson 1986). Genetic bottlenecks occur because introduced individuals represent only a sub-sample of genetic diversity in the native area. In addition, genetic drift leads to a loss of rare genotypes that occurs by chance in small founder populations. This may result in a reduced genetic diversity in the founder populations and a lower probability of population persistence (Allendorf and Lundquist 2003). Therefore, successful invasive species either have to cope with a genetic bottleneck or overcome it. Rapid evolutionary processes also occur due to the response to new selection pressures in the novel environment (Sakai et al. 2001) as well as the release of selection pressure from the old environment (Maron et al. 2004; Prati and Bossdorf 2004). The Evolution of Increased Competitive Ability (EICA) hypothesis, for instance, predicts that, compared to populations in the native area, invasive populations are selected for increased growth and reproduction at the expense of defence mechanisms against specialist herbivores or pathogens that are not present in the introduced range (Blossey and Noetzold 1995). Such evolutionary adjustments may be one possible explanation for the time lag after which many non-native species become invasive (Kowarik 1995). It is assumed that evolutionary adjustments to the new environmental conditions take place between the introduction of a species and when it begins to spread (Mooney and Cleland 2001). Finally, one important process that may cause evolutionary changes in invasive species is the hybridisation with either native relatives or other introduced species (Ellstrand and Schierenbeck 2000). Hybridisation may promote invasiveness by creating novel genotypes, fixation of heterotic genotypes, dumping of genetic load and creating high genetic variation (Ellstrand and Schierenbeck 2000). While the first three processes may directly lead to invasive genotypes, high levels of genetic variation help to overcome a genetic bottleneck and provide the raw material for the response to natural selection, and are thus necessary for adaptation to the new habitats (Blows and Hoffmann 2005; Kawecki and Ebert 2004). Rapid evolutionary responses to small-scale environmental conditions may increase the extension of the invaded area (Weber and Schmid 1998) as well as the number of invaded habitats within regions and the dominance of the species within habitats (Parker et al. 2003).

Besides genetic adaptation, phenotypic plasticity is assumed to be important for the invasion success of alien plants (Barrett and Richardson 1986). Phenotypic plasticity allows plants to express advantageous phenotypes in different environments, and thus enhances niche breath (Sultan et al. 1998). Phenotypic plasticity is generally considered as a key characteristic of colonising species (Baker 1965) and has been found to play an important role in the invasion of several species (Dybdahl and Kane 2005; Mal and Lovett-Doust 2005; Parker et al. 2003). Since phenotypic plasticity has a genetic basis and is subjected to selection, reshuffling of genomes by hybridisation can affect phenotypic plasticity (Schlichting 1986). Since high phenotypic plasticity may confer a fitness advantage in new environments, it has been suggested that invasive populations might have evolved greater plasticity than conspecific populations in the native range (Richards et al. 2006; Bossdorf et al. 2005).

1.3   Cultivated plants

Although many invasive plant species were originally imported as crops or ornamentals (see Kowarik 2004; Preston et al. 2002; Klotz et al. 2002), this pathway of introduction has been neglected for by researchers a long time. Nevertheless, genetic changes are particularly common among cultivated plants (Kowarik 2005), and cultivation may influence the invasion potential of non-native species (Kitajima et al. 2006). Cultivated plants may be particularly successful invaders because of several different mechanisms. First, ornamental or crop plants are not randomly chosen from native populations (Kitajima et al. 2006) but are selected according to beauty, habit and vigour. Consequently, there may already be a shift toward particularly fit genotypes prior to or during introduction. Second, the likelihood of spread of cultivated plants in the new area is high because of stochastic effects (Mack 2000) that appear when cultivated plants are planted in large numbers over wide areas that are protected from detrimental environmental factors (Kowarik 2005). Thereby, many propagules can disperse into adjacent habitats thus increasing the probability of successful establishment (Mack 2000). Third, evolutionary changes in cultivated plants occur due to breeding and artificial selection (Kitajima et al. 2006). It is assumed that certain traits which enhance growth and reproduction are favoured by plant breeders. Such fitness-related traits may increase the species' success not only in gardens but also in natural habitats (Kitajima et al. 2006). Furthermore, breeders often use hybridisation to enhance genetic variability for selection, and to create new genotypes with novel trait expression, combinations of genes, or increased suitability in the new habitat (Bundesverband Deutscher Pflanzenzüchter e.V. 2007). Artificial selection during introduction and plant breeding, and ubiquitous planting of the selected genotypes is assumed to enhance the invasiveness of a species (Kitajima et al. 2006; Kowarik 2005). In several cultivated plant species which have become invasive, hybridisation has already been detected, for instance in invasive *Fallopia* species (Hollingsworth et al. 1998) or invasive *Rhododendron ponticum* (Milne and Abbott 2000). In San Francisco Bay, hybrids between exotic and deliberately introduced *Spartina alterniflora* and the native *S. folios* outcompete native species (Ayres et al. 1999).

1.4   Mahonia aquifolium

Mahonia aquifolium PURSH. (NUTT.) (Berberidaceae) is an ornamental plant species that is invasive in Central Europe. The genus *Mahonia* NUTT. is treated as either distinct from the genus *Berberis* since the beginning of the twentieth century (Ahrendt 1961), or included into *Berberis* (Kim et al. 2004; Laferriere 1997). I will follow the taxonomic revision of Ahrendt (1961) in which *Mahonia* is considered as a genus distinct from *Berberis*. The genus *Mahonia* comprises fleshy-fruited evergreen shrubs with pinnate leaves and yellow flowers and is native to Asia and North America (Ahrendt 1961). Within the genus, *M. aquifolium* belongs to the section Aquifoliatae Fedde., which is native in North America. *M. aquifolium* occurs in the western states, i.e. British Columbia, California, Idaho, Montana, Washington and Oregon (Whittemore 1997) (Figure 1). The stem is simple and erect and reaches 1.80 m in height (Ahrendt 1961; Piper 1922). The leaves are shiny above and dull underneath (Piper 1922). The shrub grows in open woodland and shrubland up to 2100 m above sea level.

Mahonia aquifolium was first introduced to Europe for ornamental purposes in 1822 (Hayne 1822, cited in Kowarik 1992) and later repeatedly. The first spontaneous occurrence outside gardens was observed in 1860 after a time lag of 38 years (Kowarik 1992). Cultivated *M. aquifolium* was intensively bred, including hybridisation with related species, in particular *M. repens* (LINDL.) G.DON (Ahrendt 1961) and *M. pinnata* (Lag.) FEDDE. as is indicated by many cultivated hybrids (van de Laar 1975). Both, *M. repens* and *M. pinnata* are members of the section Aquifoliatae and are native to North America, too (Figure 1). *Mahonia repens* has a large distribution area in western and central North America that overlaps with the distribution area of *M. aquifolium*. In contrast to *M. aquifolium*, *M. repens* grows in open forests and grasslands (Whittemore 1997). It is morphologically very similar to *M. aquifolium* and some specimens can be assigned to one of the two species only with difficulties (Ahrendt 1961). *Mahonia repens* reaches only 90 cm in height and growth is often more stoloniferous than *M. aquifolium* (Ahrendt 1961). In addition its leaves are mostly dull above (Piper 1922). *M. pinnata* occurs at the coastline of the states of California, Oregon and Mexico, in exposed rocky openings in woods and shrublands (Whittemore 1997). *M. pinnata* attaines a height of 3 m (Ahrendt 1961) and has shiny leaves above and underneath. In contrast to the other two species, the first leaflets of the pinnate leaf grow near the base of the petiole (Munz 1959).

The breeding of *M. aquifolium* and the two related species has resulted in many cultivars (Houtman et al. 2004; van de Laar 1975) (*M. x decumbens* = *M. aquifolium* x *M. repens*; *M. x wagneri* = *M. aquifolium* x *M. pinnata)* (Figure 2), that were frequently planted in gardens, parks and along roads (Figure 3). *M. aquifolium* produces many fleshy fruits that are eaten by birds, which disperse seeds into adjacent vegetation. Today, the species is spreading and invading anthropogenic and natural habitats in Central Europe (Kowarik 1992). Like the whole genus, *M. aquifolium* is characterised as an outbreeding species (Burd 1994). The species is self-incompatible because self-pollination rarely results in fruit production (Monzingo 1987, H. Auge unpublished data). The invasive populations disperse not only by seeds but also by stolons and stem layers (Auge and Brandl 1997), which is less known from native *M. aquifolium* but commonly from *M. repens* (Ahrendt 1961) and also found in some cultivars (Houtman et al. 2004). However, sexual reproduction seems to play a major role for regional as well as for local spread (Auge 1997). *Mahonia* individuals usually flower beginning in the third year (personal observation) from April to June and produce multitudes of berries in September and October, which remain attached to the plant until winter (Zeitlhöfler 2002). Invasive populations grow either continuously or in patches (Figure 4). It is assumed that invasive populations have descended from cultivars that are mostly hybrids (van de Laar 1975; Ahrendt 1961). In consequence, it is likely that invasive populations also consist of hybrids. Since the identity of invasive *M. aquifolium* is not known, I will use the term invasive *Mahonia* populations instead of invasive *M. aquifolium* from here on.

Invasive *Mahonia* appears to be well suited for a case study to investigate the importance of plant breeding for invasion success, because (1) it is one of the most successful alien shrubs in central and eastern Germany (Kowarik 1992) (2) it was introduced as an ornamental plant, (3) it was modified by plant breeders, and (4) the propagule pressure of *Mahonia* to natural habitats is high, because it is planted in great numbers in Central Europe.

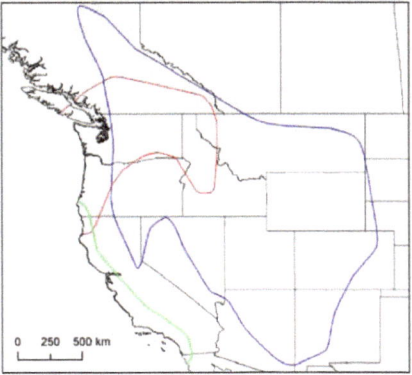

Figure 1: Map of the distribution areas of the three native species after Whittemore (1997) *M. aquifolium* (red), *M. repens* (blue) and *M. pinnata* (green) in western North America.

Figure 2: Cultivated individuals of *Mahonia* are very diverse in height, growth form, morphology, colour and quantity of flowers. a) *M. wagneri* 'Pinnacle' b) *M. aquifolium* 'Hans-Karl Möhring' c) *M. aquifolium* 'Undulata' d) *M. aquifolium* 'Hasting Elegant' e) *M. repens* f) *M. aquifolium* 'Brilliant'.

Figure 3: Cultivated *Mahonia* plants along a street in the city of Halle in Saxony-Anhalt (Germany).

Figure 4: Two invasive *Mahonia* populations in a pine forest of Dübener Heide in Saxony-Anhalt (Germany). The left population grows continuously throughout the forest, the right population grows in patches.

1.5   Thesis objectives

The aim of the studies presented in this thesis is to examine the importance of evolutionary changes, especially due to plant breeding, in the invasion success of *Mahonia* in Germany. I conducted three studies. (1) I compared native and invasive *Mahonia* populations using neutral molecular markers. Due to the markers' neutrality, different selection pressures could be excluded as a reason for differences in allele frequencies, thus allowing the investigation of genetic bottlenecks and hybridisation. (2) I compared native and invasive *Mahonia* populations by using approaches of quantitative genetics to investigate whether evolutionary changes in ecologically relevant traits have occurred. (3) I studied the importance of local genetic adaptation for the successful colonisation of different habitats in the invasive area.

1.5.1   Differences in neutral molecular markers (chapter 2 and chapter 3)
The first study aimed at detecting whether genetic changes resulting from a genetic bottleneck effect or hybridisation happened during the invasion of *Mahonia* in Central Europe. The genetic origin of the invasive populations was investigated by neutral genetic DNA markers which were not likely to be affected by natural selection. I used microsatellite markers, the isolation and characterisation of which are described in chapter 2. Microsatellite markers are DNA units of a few base pairs that are repeated in variable numbers and often have a multitude of alleles that vary in the number of repeats (Queller et al. 1993). Individuals, populations and species can be compared by their allele frequencies and, thus, microsatellite markers are a powerful tool for several questions in population biology, and in particular provide information about kinship (Queller et al. 1993). I isolated ten markers which were polymorphic in *M. aquifolium*, *M. repens* and *M. pinnata*. Eight of these ten markers were used in the subsequent comparison of native populations of *M. aquifolium*, *M. repens*, *M. pinnata* with invasive populations and cultivars (chapter 3). In this study the genetic relationship between cultivars, native species and invasives was investigated (table 1).

I did not find reduced genetic diversity in invasive *Mahonia* populations compared to native populations, indicating that invasive populations are not chara- cterised by a genetic bottleneck. However, invasive populations are genetically different from native populations of all three species. Clustering of native species separated *M. repens* into two groups that reflected the geographical pattern of sampling. *M. aquifolium* was closely related to the *M. repens* populations that were collected at the northern end of the *M. repens* range. This close relationship suggests that the species are either connected by current gene flow in the overlapping distribution area or indicates an intermediate state of ongoing speciation. Furthermore, the close relationship of *M. aquifolium* and *M. repens* complicated the analyses of relationships between invasives, cultivars and native species; in particular, hybridisation was difficult to detect. Cultivars had a higher probability to be members of *M. pinnata* than members of *M. repens*. This difference may be an effect of early hybridisation of *M. aquifolium* and *M. pinnata* (by 1930) and the fact that old cultivars were used for breeding of new cultivars. The finding of a high probability of some cultivars to be *M. pinnata* nevertheless they are not referred to *M. pinnata* or *M. pinnata* hybrids by breeders, affirms this suggestion. In contrast, hybridisation of *M. aquifolium* with *M. repens* was carried out later by plant breeders and cultivars were, therefore, relatively young. However, all invasive populations had only a low probability to be composed of *M. pinnata* and *M. repens*, indicating that those cultivars with a high probability to be *M. pinnata* were not aggressive invaders of natural habitats.

According to the results of the molecular genetic analyses, I assume that the differentiation of invasive *Mahonia* and native *M. aquifolium* is a result of the inter-action of different processes that happened during the invasion of *Mahonia*, including restriction of gene pools, genetic drift, artificial selection and hybridisation. Thus, although a significantly reduced genetic diversity could not be detected in invasive *Mahonia* populations compared to native populations, it is likely that selective intro-duction of certain genotypes reduced genetic variability. Probably, genetic variability was enhanced by breeding and hybridisation after introduction. Simultaneously, genetic makeup was likely directly changed by artificial selection, which may cause a bias in invasive founder populations. I assume that the genetic makeup changed another time in the stage of secondary release, because some cultivars were found to be more invasive than others. Thus, it is likely that plant breeding facilitated the invasion success of *Mahonia* by enhancing genetic variability and by generating characteristics that increase invasiveness of certain cultivars.

1.5.2   Genetic differences in ecologically relevant traits (chapter 4)

Microsatellite markers are neutral molecular genetic markers that measure genetic variation which is not associated with any measurable phenotypic variation (Griffiths et al. 1997) and, therefore, are not subjected to natural selection. Thus, allele frequencies of such neutral markers provide information about hybridisation and the genetic variation of invasive populations, but do not inform about the ecological relevance of genetic changes. An ecologically relevant change of invasive populations is, for instance, that these plants often grow more vigorously and produce more seeds than their relatives in the native area (Crawley 1987). This pattern has been traced back to genetic changes by some authors (Bossdorf et al. 2005; Blossey and Noetzold 1995). Such genetically based phenotypic characteristics usually vary continuously and are influenced by several genes (Griffiths et al. 1997). They are referred to as quantitative genetic traits and since they are often associated with fitness differences among phenotypes, quantitative genetic traits are subjected to natural selection. Sufficiently large heritable phenotypic variation is a prerequisite for a populations' response to natural selection, and thus for adaptive evolution (Lynch and Walsh 1998). If genetic variation was reduced during introduction, adaptive evolution in invasive populations would be impeded by this bottleneck effect (Allendorf and Lundquist 2003). However, multiple introductions and hybridisation can enhance variation, making adaptations to the new environment possible. Comparing offspring of several maternal plants allows quantifying the genetic component of phenotypic variation in ecologically important traits within populations.

I investigated quantitative genetic traits of two native *Mahonia* species, *M. aquifolium* and *M. repens*, and invasive populations in the fourth chapter of my thesis. In that experiment, invasive *Mahonia* populations and native *Mahonia* populations were grown together in the same environment in a common garden. Thus, most of the phenotypic variation among them can be attributed to genetic differentiation (Griffiths et al. 1997). In addition to different taxa and different populations within taxa, I compared offspring of several maternal plants within populations and compared the heritable variation of native and invasive populations. I found that invasive *Mahonia* populations showed a genetically based, strongly increased plant size compared to the native species (table 1). More vigorous plant growth has been recognised as a general phenomenon in plant invasions (Crawley 1987) and has been traced back to genetic changes by some authors (Brown and Eckert 2005; Willis and Blossey 1999; Blossey and Noetzold 1995). I discuss the evolutionary changes of invasive *Mahonia* populations in terms of plant breeding involving selective introduction, hybridisation and artificial selection. I suspect that early hybrid genotypes formed the raw material, which was then used by plant breeders to select for particularly vigorous growth. In addition, I found no significant differences in the magnitude of heritable variation between invasive and native *Mahonia* populations (table 1) indicating that multiple introduction and hybridisation have counteracted a genetic bottleneck.

1.5.3 Local adaptation (chapter 5)

Since we found no reductions in heritable genetic variation in invasive *Mahonia* populations, adaptation to different environmental conditions may be a possible mechanism for invasion success of *Mahonia*. In this part of the thesis, I investigated whether local adaptation to different environments has occurred in invasive *Mahonia*. These populations invade different habitats, and local genetic adaptation is, apart from phenotypic plasticity, suggested to be a mechanism permitting survival under different abiotic and biotic conditions. Local adaptation is common in plants (Linhart and Grant 1996) and is a result of divergent selection pressures in different habitats. In invasive plants, the ability to rapidly adapt to different habitats may increase both the number of invaded habitats within regions and the dominance within habitats (Parker et al. 2003). The ability to adapt is therefore regarded as a key feature of successful invaders (Sakai et al. 2001). The response to selection of a trait is proportional to its heritable variation, but genetic variation of invasive populations is often reduced due to a genetic bottleneck after introduction. Invasive *Mahonia* populations, however, show great variation in quantitative traits (Auge 1997; Ahrendt 1961) that are not reduced in comparison to native populations (chapter 4). In addition, I did not find a genetic bottleneck in invasive *Mahonia* populations using molecular markers (chapter 3). Thus, phenotypic and genetic variation provide the ability for local adaptation in invasive *Mahonia* populations.

Whether invasive populations are adapted to different habitats was the question of this fifth chapter. I used two reciprocal transplant experiments in which I transplanted seedlings of invasive *Mahonia* populations to five different soil types or sites, respectively. Despite the large phenotypic and genetic variation in invasive *Mahonia* populations and the differences in environmental conditions between the five sites, which is the precondition for local adaptation, I did not find any evidence for local adaptation (table 1). Growth strongly depended on the soil and the sites, and seedlings of all populations showed the best growth in the same soil or at the same site, respectively. The "best" soil was not taken from the "best" site, indicating that habitat quality to *Mahonia* seedlings is not determined by soil quality alone. There are different possibilities for I found no evidence for local adaptation. First, although I detected high phenotypic and genetic variation in invasive *Mahonia* populations in the previously described studies, I found no differences in response of the seed families to the different soils in this study. Thus, a lack of heritable variation may have prevented a response to divergent selection and thus genetic adaptation. Second, high gene flow between different invasive populations and between these populations and cultivated individuals may counteract an adaptation. Third, the time since *Mahonia* has invaded the habitats (ca. 40-80 years ago) may be too short to detect local adaptation. Last but not least, other life cycle stages aside from the seedling stage may play a more important role in local adaptation of invasive *Mahonia*.

Table 1: Investigated objectives and detected results, sorted according to the chapters in which they are discussed in this thesis.

Investigated objectives	Results
Differences in neutral molecular markers (chapter 3)	
Genetic bottleneck in invasive populations	No genetic bottleneck was detectible in invasive *Mahonia* populations.
Genetic differences between invasive and native *Mahonia* populations	Invasive *Mahonia* populations are in allele identity respectively in allele frequency different to native populations.
Relatedness between cultivars and invasive and native *Mahonia* populations	Cultivars could often not be assigned genetically to the species they are referred to. Some cultivars were genetically more similar to invasive *Mahonia* populations than others.
Hybrid origin of invasive populations	No hybridisation origin could be detect in molecular data
Genetic differences in ecological relevant traits (chapter 4)	
Differences in quantitative genetic traits	Invasive populations grew more vigorously than native *Mahonia* populations.
Differences in heritable variation of quantitative genetic traits	Heritable variation was similar in invasive and native *Mahonia* populations.
Local adaptation (chapter 5)	
Local adaptation of invasive *Mahonia* populations to soil conditions	Invasive *Mahonia* populations are not local adapted their home soil
Local adaptation of invasive *Mahonia* populations to habitats	Invasive *Mahonia* populations are not local adapted their home soil

1.6   General Discussion

The results of my studies on the evolution of invasive *Mahonia* are that (1) invasive *Mahonia* populations are genetically different from all three native species, (2) invasive populations display similar genetic diversity compared to native species, indicating that no genetic bottleneck occurred in invasive populations, (3) invasive populations grew more vigorously than native *M. aquifolium* and native *M. repens* in a common garden, and (4) in spite of high phenotypic and genetic variation there was no evidence for local genetic adaptation to different environments.

1.6.1   Evolutionary changes occurred in invasive Mahonia

The molecular genetic investigations as well as the quantitative genetic experiments indicate evolutionary changes in invasive *Mahonia* populations. Invasive *Mahonia* were genetically different from native populations and displayed more vigorous growth. Thus, I suggest that rapid evolution occurred during the invasion of *Mahonia* in Central Europe. Several studies have showed evolutionary changes in other invasive plant species (Erfmeier and Bruelheide 2004; Jakobs et al. 2004; Prati and Bossdorf 2004). Two of these three examples investigated cultivated plants, but did not explicitly explore breeding and cultivation as factors contributing to the successful invasion. Although the invasion success of cultivated plants is high (Klotz et al. 2002), the importance of breeding and cultivation has generally been marginalized in the study of biological invasions (but see Kitajima et al. 2006). Instead, most studies investigate either genetic bottlenecks or the EICA hypothesis as a reason for genetic changes in invasive populations (Bossdorf et al. 2005). However, neither of these two possibilities provides a plausible explanation for the invasion success of *Mahonia*. Genetic variation, both in terms of neutral molecular markers and phenotypic traits, was similar in native and invasive populations indicating an invasion without a genetic bottleneck. Furthermore, invasion success cannot be attributed to the EICA hypothesis, because no release from specialist herbivores is known in *Mahonia* (Soldaat and Auge 1998; Auge et al. 1997; M. Brändle, C. Belle and R. Brandl, unpublished data).

1.6.2   Plant cultivation facilitated the invasion of Mahonia

I investigated the role of plant breeding in facilitating the invasion of *Mahonia* in different ways. Certain genotypes may have been selectively introduced into Central Europe. This would result in reduced genetic diversity, which I did not find in the invasive *Mahonia* populations. Therefore, my results support the finding by Bossdorf et al. (2005) that most plant invasions are not associated with a genetic bottleneck and multiple introduction of invasive plants seem to be the rule. In addition to multiple introductions, hybridisation may counteract a harmful genetic bottleneck. Hybridisation is known to create novel genotypes and high genetic variation, which may facilitate genetic adaptation to new environmental conditions (Ellstrand and Schierenbeck 2000). *M. aquifolium* was traceably hybridised with the related species *M. repens* and *M. pinnata*, which resulted in many hybrid cultivars (van de Laar 1975). Since *M. aquifolium* and *M. repens* could not be separated by microsatellite markers, detection of hybrids between them was difficult. My results reveal that *M. aquifolium* and *M. repens* cannot be considered as two clearly separated species and, thus, contribute to the discussion about the systematic of the two species (Piper 1922; Piper 1906).

In addition to the change in genetic makeup of cultivated *Mahonia* plants, plant breeding facilitated the invasion by enhancing propagule pressure due to large quantities of planted individuals of various cultivars and casual hybrids. Propagule pressure has generally been shown to be one of the few factors that can be identified in determining invasion success (Rejmanek 2000). I found very few cities in Germany where *Mahonia* was not planted in gardens and along streets (personal observation). *Mahonia* seeds are bird dispersed and in that way, many seeds can reach adjacent natural habitats (Auge and Brandl 1997). Seed dispersal not only contributes to colonisation of new sites but may also lead to permanent gene flow into established *Mahonia* populations. This gene flow may be one reason for the high genetic diversity observed in invasive populations, and may also hinder local adaptation (Kawecki and Ebert 2004). Furthermore, the genetic diversity of invasive populations is an indication of the genetic diversity of cultivated individuals. Although *Mahonia* cultivars are propagated clonally (personal communication with several breeders), my genetic analyses detected a high genetic diversity of cultivars. Nevertheless, I show that not all cultivars are similarly successful in invading natural habitats. In particular, those cultivars in which I detected genetic material of *M. pinnata* seem to be less aggressive invaders, which can be either a mass effect or a breeding effect. Dehnen-Schmutz et al. (2007) demonstrated that those ornamentals which are more available at the market more frequently escape into natural habitats. Of course, there are also *Mahonia* cultivars that are more available at the market, and others that are less so. In an inquiry of 14 nurseries, I found that 11 nurseries sold *M. aquifolium* (*M. aquifolium* without a cultivar's name was grown from seeds of diverse origins, personal communication with breeders), 11 sold *M. aquifolium* 'Apollo', four nurseries sold *M. aquifolium* 'Smaragd', three sold *M. aquifolium* 'Atropurpurea', and five nurseries sold one or few other cultivars of *M. aquifolium* and hybrids. Thus, the genetic makeup of invasive populations can be a result of the fact that *M. aquifolium* was planted more often than other cultivars that contain also genetic material from the two other species, *M. repens* and *M. pinnata*. In addition to this bias in planting frequency of the different cultivars, they may vary greatly in invasiveness, which may be another reason for finding only a sub-sample of the genetic material of cultivars in invasive populations.

1.6.3   Evolved phenotypic plasticity as a reason for invasiveness

The results of the molecular genetic analyses show that the different invasive populations are very similar to each other. I found that they were weakly structured and not isolated by distance. These results indicate that all invasive *Mahonia* populations were founded by individuals with rather similar genetic material. I suggest that plant breeders have perhaps not created *Mahonia* cultivars that differ in their success under different environmental conditions, but created such cultivars that are fit in all habitats and, thus, fall in the category of Jack-of-all-trades (Richards et al. 2006). This kind of phenotypic plasticity is related to the concept of "general purpose genotypes" (Baker 1965) and describes plants, which are able to maintain their fitness across different environments (Richards et al. 2006). In contrast, there is another category, referred to as Master-of-some, which describes the ability of a species to increase fitness in favourable environments. The concept of phenotypic plasticity has been shown to be important in several invasive species (Dybdahl and Kane 2005; Mal and Lovett-Doust

2005; Parker et al. 2003). My study on local adaptation supports the view that invasive *Mahonia* populations evolved highly plastic genotypes rather than genotypes that are genetically adapted to specific environments. The results of this study indicate a high phenotypic plasticity of *Mahonia* seedlings in response to the different habitats. Despite the large genetic diversity in invasive *Mahonia* populations, which is a prerequisite for response to natural selection, I did not find local adaptation to different habitats in invasive *Mahonia* seedlings. Moreover, seedlings of all populations grew best in the same habitat, and worst in the same other habitat, a pattern that indeed follows the Master-of-some scenario. Richards et al (2006) described that invaders might combine the Jack-of-all-trades and Master-of-some situation and are a Jack-and-master. Possibly that is true for invasive *Mahonia*. Since phenotypic plasticity is a trait that is genetically based and subjected to selection (Schlichting 1986), the high genetic diversity in *Mahonia* could have fostered the selection of plastic genotypes. Furthermore, the selection of plastic genotypes is a benefit to cultivated plant species, because plastic cultivars can be planted in different environmental conditions, and thus breeders could select plastic individuals. Whether phenotypic plasticity actually plays a role for the invasion success of *Mahonia* should be investigated in further studies. Nevertheless, my results give hints that phenotypic plasticity may play a role in *Mahonia* invasions. The common garden experiment revealed indeed no difference between native and invasive populations in the response of growth-related traits to shade. However, phenotypic plasticity refers to the response of a certain trait to certain environments, and different environments can therefore yield different results (Richards et al. 2006). Whether evolution of high phenotypic plasticity, or local genetic adaptation in life stages other than seedlings, or both, contributes to invasion success is not yet answerable.

1.6.4   Importance of evolution, cultivation, and breeding in studies of plant invasions

The importance of evolutionary changes for the invasiveness of plant species has been shown in several studies before (reviewed in Bossdorf et al. 2005). I provide evidence that such evolutionary changes have also occurred in invasive *Mahonia* populations. Only a few studies have explored breeding and cultivation as a possible cause of invasiveness (but see Kitajima et al. 2006), although many invasive plants are cultivated plants (Klotz et al. 2002). It was not possible to provide unequivocal evidence that hybridisation of *Mahonia* resulted in invasive genotypes. Nevertheless, my results indicate at least that the high genetic variation, which presumably is a result of plant breeding, fostered the *Mahonia* invasion, and that evolution of phenotypic plasticity may be an important factor. This study therefore contributes to a general assessment of the role of evolution in plant invasions. It demonstrates that the role of plant breeding and cultivation should not be neglected when searching for characteristics of successful invaders, and requests that scientists, plant breeders and gardeners cooperate in order to mitigate current and to prevent future plant invasions.

References

Ahrendt, L.W.A. (1961) *Berberis* and *Mahonia*. A taxonomic revision. *Journal of the Linnean Society of London, Botany,* **57**: 1-410.

Allendorf, F.W. and Lundquist, L.L. (2003) Introduction: Population Biology, Evolution, and Control of Invasive Species. *Conservation Biology,* **17**: 24-30.

Armstrong, S. (1995) Rare plants protect Cape's water supplies. *New Scientist,* **11**: 8.

Auge, H. (1997) Biologische Invasionen: Das Beispiel *Mahonia aquifolium*. In: *Regeneration und nachhaltige Landnutzung - Konzepte für belastete Regionen* (R. Feldmann, K. Henle, H. Auge, J. Flachowsky, S. Klotz and R. Kroenert, eds), pp. 124-129 Berlin: Springer Verlag.

Auge, H. and Brandl, R. (1997) Seedling recruitment in the invasive clonal shrub, *Mahonia aquifolium* Pursh (Nutt.). *Oecologia,* **110**: 205-211.

Auge, H., Brandl, R. and Fussy, M. (1997) Phenotypic variation, herbivory and fungal infection in the clonal shrub *Mahonia aquifolium* (Berberidaceae). *Mitteldeutsche Gesellschaft für Allgemeine und Angewandte Entomology,* **11**: 747-750.

Auge, H., Klotz, S., Prati, D. and Brandl, R. (2001) Die Dynamik von Pflanzeninvasionen: ein Spiegel grundlegender ökologischer und evolutionsbiologischer Prozesse. *Rundgespräche der Kommission für Ökologie,* **22**: 41-58.

Ayres, D.R., Garcia-Rossi, D., Davis, H.G. and Strong, D.R. (1999) Extent and degree of hybridization between exotic (*Spartina alterniflora*) and native (*S. foliosa*) cordgrass (Poaceae) in California, USA determined by random amplified polymorphic DNA (RAPDs). *Molecular Ecology,* **8**: 1179-1186.

Baker, H.G. (1965) Characteristics and Modes of Origin of Weeds. In: *The Genetics of Colonizing Species: Proceedings of the First International Union of Biological Sciences Symposia on General Biology* (H.G. Baker and G.L. Stebbins, eds), pp. 147-168 New York: Academic Press Inc.

Barrett, S.C.H., and Richardson, B.J. (1986) Genetic attributes of invading species. In: *Ecology of biological invasions* (R.H. Groves and J.J. Burdon, eds), pp. 21-33 Cambridge: Cambridge University Press.

Belnap, J., Phillips, S.L., Sherrod, S.K. and Moldenke, A. (2005) Soil biota can change after exotic plant invasion: Does this affect ecosystem processes? *Ecology,* **86**: 3007-3017.

Blossey, B. and Noetzold, R. (1995) Evolution of increased competitive ability in invasive nonindigenous plants: A hypothesis. *Journal of Ecology,* **83**: 887-889.

Blows, M.W. and Hoffmann, A.A. (2005) A reassessment of genetic limits to evolutionary change. *Ecology,* **86**: 1371-1384.

Bossdorf, O., Auge, H., Lafuma, L., Rogers, W.E., Siemann, E. and Prati, D. (2005) Phenotypic and genetic differentiation between native and introduced plant populations. *Oecologia,* **144**: 1-11.

Brown, J.S. and Eckert, C.G. (2005) Evolutionary increase in sexual and clonal reproductive capacity during biological invasion in an aquatic plant *Butomus umbellatus* (Butomaceae). *American Journal of Botany,* **92**: 495-502.

Bundesverband Deutscher Pflanzenzüchter e.V. (2007) Züchtung. http://www.bdp-online.de/index.php?menu=3.

Burd, M. (1994) Bateman principle and plant reproduction - the role of pollen limitation in fruit and seed set. *Botanical Review,* **60**: 83-139.

Callaway, R.M. and Aschehoug, E.T. (2000) Invasive Plants Versus Their New and Old Neighbors: A Mechanism for Exotic Invasion. *Science,* **290**: 521-523.

Caroll, S.P. and Dingle, H. (1996) The biology of post-invasion events. *Biological Conservation,* **78**: 207-214.

Crawley, M.J. (1987) What makes a community invasible? In: *Colonization, succession and stability* (A.J. Gray, M.J. Crawley and P.J. Edwards, eds), pp. 429-453 Oxford: Blackwell Scientific Publications.

Dehnen-Schmutz, K., Touza, J., Perrings, C. and Williamson, M. (2007) The horticultural trade and ornamental plant invasions in Britain. *Conservation Biology,* **21**: 224-231.

Drake, J.A., Mooney, H.A., di Castri, F., Groves, R.M., Kruger, F.J., Rejmanek, M. and Williamson, M. (1989) Biological Invasions: A Global Perspective, Chichester: John Wiley & Sons.

Dukes, J.S. and Mooney, H.A. (1999) Does global change increase the success of biological invaders. *Trends in Ecology & Evolution*, **14**: 135-139.

Dybdahl, M.F. and Kane, S.L. (2005) Adaptation vs. phenotypic plasticity in the success of a clonal invader. *Ecology*, **86**: 1592-1601.

Ellstrand, N.C. and Schierenbeck, K.A. (2000) Hybridization as a stimulus for the evolution of invasiveness in plants. *Proceedings of the National Academy of Sciences of the United States of America*, **97**: 7043-7050.

Erfmeier, A. and Bruelheide, H. (2004) Comparison of native and invasive *Rhododendron ponticum* populations: Growth, reproduction and morphology under field conditions. *Flora*, **199**: 120-133.

Griffiths, A.J.F., Miller, J.H., Suzuki, D.T., Lewontin, R.C. and Gelbart, W.M. (1997) An introduction to genetic analysis, New York: W. H. Freeman and Company.

Hollingsworth, M.L., Hollingsworth, P.M., Jenkins, G.I., Bailey, J.P. and Ferris, C. (1998) The use of molecular markers to study patterns of genotypic diversity in some invasive alien *Fallopia* spp. (Polygonaceae). *Molecular Ecology*, **7**: 1681-1691.

Houtman, R.T., Kraan, K.J. and Kromhout, H. (2004) *Mahonia aquifolium, M. repens, M. x wagneri* en hybriden. *Dendroflora*, **41**: 42-69.

Jakobs, G., Weber, E. and Edwards, P.J. (2004) Introduced plants of the invasive *Solidago gigantea* (Asteraceae) are larger and grow denser than conspecifics in the native range. *Diversity and Distributions*, **10**: 11-19.

Kawecki, T.J. and Ebert, D. (2004) Conceptual issues in local adaptation. *Ecology Letters*, **7**: 1225-1241.

Kim, Y.-D., Kim, S.-H. and Landrum, L.R. (2004) Taxonomic and phytogeographic implications from ITS phylogeny in *Berberis* (Berberidaceae). *Journal of Plant Research*, **117**: 175-182.

Kitajima, K., Fox, A.M., Sato, T. and Nagamatsu, D. (2006) Cultivar selection prior to introduction may increase invasiveness: evidence from *Ardisia crenata. Biological Invasions*, **8**: 1471-1482.

Klotz, S., Kühn, Ingolf, and Durka, Walter. (2002) BIOLFLOR - Eine Datenbank mit biologisch-ökologischen Merkmalen zur Flora von Deutschland. *Schriftenreihe für Vegetationskunde*, Vol. 38, Bonn - Bad Godesberg: Bundesamt für Naturschutz.

Kowarik, I. (1992) Einführung und Ausbreitung nichteinheimischer Gehölzarten in Berlin und Brandenburg. *Verhandlungen Botanischer Vereine Berlin Brandenburg,* **3**: 1-188.

Kowarik, I. (1995) Time lags in biological invasions with regard to the success and failure of alien species. In: *Plant Invasions - General Aspects and Special Problems* (P. Pysek, K. Prach, M. Rejmanek and M. Wade, eds), pp. 15-38 Amsterdam: Academic Publishing.

Kowarik, I. (2004) Biologische Invasionen in Deutschland: zur Rolle nichteinheimischer Pflanzen. In: *Biologische Invasionen. Herausforderung zum Handeln?* (I. Kowarik and U. Starfinger, eds), pp. 5-24

Kowarik, I. (2005) Urban ornamentals escaped from cultivation. In: *Crop Ferality and Volunteerism* (J. Gressel, eds), pp. 97-121 Boca Raton: CRC Press.

Laferriere, J. (1997) Transfer of specific and infraspecific taxa from *Mahonia* to *Berberis* (Berberidaceae). *Botanicheskii Zhurnal,* **82**: 95-99.

Linhart, Y.B. and Grant, M.C. (1996) Evolutionary significance of local genetic differentiation in plants. *Annual Review of Ecology and Systematics,* **27**: 237-277.

Lynch, M. and Walsh, B. (1998) Genetics and Analysis of Quantitative Traits, Sunderland: Sinauer Associates, Inc. Publishers.

Mack, R.N. (2000) Cultivation fosters plant naturalization by reducing environmental stochasticity. *Biological Invasions,* **2**: 111-122.

Mack, R.N., Simberloff, D., Lonsdale, W.M., Evans, H., Clout, M. and Bazzaz, F.A. (2000) Biotic invasions: Causes, epidemiology, global consequences, and control. *Ecological Applications,* **10**: 689-710.

Mal, T.K. and Lovett-Doust, J. (2005) Phenotypic plasticity in vegetative and reproductive traits in an invasive weed, *Lythrum salicaria* (Lythraceae), in response to soil moisture. *American Journal of Botany,* **92**: 819-825.

Maron, J.L., Vila, M., Bommarco, R., Elmendorf, S. and Beardsley, P. (2004) Rapid evolution of an invasive plant. *Ecological Monographs,* **74**: 261-280.

Millenium Ecosystem Assessment. (2005) Ecosystems and Human Well-being: Biodiversity Synthesis., Washington, DC.: World Resources Institute.

Milne, R.I. and Abbott, R.J. (2000) Origin and evolution of invasive naturalized material of *Rhododendron ponticum* L. in the British Isles. *Molecular Ecology,* **9**: 541-556.

Monzingo, H.N. (1987) Shrubs of the Great Basin, Reno: University of Nevada Press.

Mooney, H.A. and Cleland, E.E. (2001) The evolutionary impact of invasive species. *Proceedings of the National Academy of Sciences of the United States of America*, **98**: 5446-5451.

Munz, P. (1959) A California Flora, Berkeley: University of California Press.

Parker, I.M., Rodriguez, J. and Loik, M.E. (2003) An evolutionary approach to understanding the biology of invasions: Local adaptation and general-purpose genotypes in the weed *Verbascum thapsus*. *Conservation Biology*, **17**: 59-72.

Pimentel, D., Lach, L., Zuniga, R. and Morrison, D. (2000) Environmental and Economic Costs of Nonindigenous Species in the United States. *BioScience*, **50**: 53-65.

Piper, C.V. (1906) Flora of the state of Washington. *Contributions from the United States National Herbarium*, **11**: 282-283.

Piper, C.V. (1922) The identification of *Berberis aquifolium* and *Berberis repens*. *Contributions from the United States National Herbarium*, **20**: 437-451.

Prati, D., and Bossdorf, O. (2004) A comparison of native and introduced populations of the South African Ragwort *Senecio inaequidens* DC. in the field. In: *Results of the worldwide ecological studies* (S.W. Breckle, B. Schweizer and A. Fangmeier, eds), pp. 353-359 Stuttgart: Verlag Günter Heimbach.

Preston, C.D., Telfer, M.G., Arnold, H.R., Carey, P.D., Cooper, J.M., Dines, T.D., Hill, M.O., Pearman, D.A., Roy, D.B. and Smart, S.M. (2002) The changing flora of the UK, London: DEFRA: Oxford University Press.

Pysek, P. (1990) On the terminology used in plant invasion studies. In: *Plant Invasions - General Aspects and Special Problems* (P. Pysek, K. Prach, M. Rejmanek and M. Wade, eds), pp. 71-81 Amsterdam: SPB Academic Publ.

Queller, D.C., Strassmann, J.E. and Hughes, C.R. (1993) Microsatellites and kinship. *Trends in Ecology & Evolution*, **8**: 285-288.

Reinhardt, F, Herle, M., Abbott, Richard J., Bastiansen, F, and Streit, B. (2003) Economic impact of the spread of alien species in Germany, Berlin (Germany): Umwelbundesamt.

Rejmanek, M. (2000) Invasive plants: approaches and predictions. *Austral Ecology*, **25**: 497-506.

Rejmanek, M. (1996) A theory of seed plant invasiveness: The first sketch. *Biological Conservation*, **78**: 171-181.

Richards, C.L., Bossdorf, O., Muth, N.Z., Gurevitch, J. and Pigliucci, M. (2006) Jack of all trades, master of some? On the role of phenotypic plasticity in plant invasions. *Ecology Letters*, **9**: 981-993.

Richardson, D.M., Pysek, P., Rejmanek, M., Barbour, M.G., Panetta, F.D. and West, C.J. (2000) Naturalization and invasion of alien plants: concepts and definitions. *Diversity and Distributions*, 93-107.

Sakai, A.K., Allendorf, F.W., Holt, J.S., Lodge, D.M., Molofsky, J., With, K.A., Baughman, S., Cabin, R.J., Cohen, J.E., Ellstrand, N.C., McCauley, D.E., O'Neil, P., Parker, I.M., Thompson, J.N. and Weller, S.G. (2001) The population biology of invasive species. *Annual Review of Ecology and Systematics*, **32**: 305-332.

Schlichting, C.D. (1986) The evolution of phenotypic plasticity in plants. *Annual Review of Ecology and Systematics*, **17**: 667-693.

Soldaat, L.L., and Auge, H. (1998) Interactions between an invasive plant, *Mahonia aquifolium*, and a native phytophagous insect, *Rhagoletis meigenii*. In: *Plant Invasions: Ecological mechanisms and human responses* (U. Starfinger, K. Edwards, I. Kowarik and M. Williamson, eds), pp. 347-360 Leiden: Backhuys Publishers.

Sperry, L.J., Belnap, J. and Evans, R.D. (2006) *Bromus tectorum* invasion alters nitrogen dynamics in an undisturbed arid grassland ecosystem. *Ecology*, **87**: 603-615.

Sultan, S.E., Wilczek, A.M., Hann, S.D. and Brosi, B.J. (1998) Contrasting ecological breadth of co-occuring annual *Polygonum* species. *Journal of Ecology*, **86**: 363-383.

van de Laar, H.J. (1975) *Mahonia en Mahoberberis. Dendroflora*, **11/12**: 19-33.

Weber, E. and Schmid, B. (1998) Latitudinal population differentiation in two species of *Solidago* (Asteraceae) introduced into Europe. *American Journal of Botany*, **85**: 1110-1121.

Whittemore, A.T. (1997) *Berberis*. In: *Flora of North America* pp. 276-286 New York, Oxford: Oxford University Press.

Williamson, M. (1996) Biological invasion, London: Chapman & Hall.

Williamson, M. and Fitter, A. (1996) The varying success of invaders. *Ecology*, **77**: 1661-1666.

Willis, A.J. and Blossey, B. (1999) Benign environments do not explain the increased vigour of non-indigenous plants: a cross-continental transplant experiment. *Biocontrol Science and Technology*, **9**: 567-577.

Zeitlhöfler, Andreas (2002) *Mahonia aquifolium* - Die Gemeine Mahonie. http://www.garteninfos.de/wildobst/Dipl4-11.html.

Chapter 2

Isolation and characterisation of microsatellite markers in the invasive shrub *Mahonia aquifolium* (Berberidaceae) and their applicability in related species

In cooperation with Walter Durka

2.1   Abstract

Microsatellite loci were isolated from a *Mahonia aquifolium* cultivar. We describe the variability of ten loci in invasive European and native North American *M. aquifolium* and their trans species amplification in native *M. repens* and *M. pinnata* from North America and one species of the related genus *Berberis* (*B. vulgaris*), native to Europe. The markers should be useful to reveal the genetic origin of invasive *Mahonia* populations and differences in the genetic make up between invasive and native populations.

2.2   Introduction

Mahonia aquifolium (PURSH) NUTT. (Berberidaceae) is a diploid (2n = 28) evergreen shrub, native to western North America. It is a successful neophyte in Central Europe (Kowarik 1992) and invades a wide range of habitats from calcareous mixed forests and xerothermic shrub vegetation to pine forests on sandy soils. The invasive populations reproduce either by seedlings or by root sprouts and stem layers. *M. aquifolium* was introduced in 1822 for ornamental purposes to Central Europe (Hayne 1822, cited in Kowarik 1992). It was supposed that invasive *Mahonia* populations mostly originated from garden plants and consist largely of hybrids between the related species *M. aquifolium* and *M. repens* (LINDL.) G.DON (Ahrendt 1961). As there are many cultivated hybrids of *M. aquifolium* with *M. pinnata* (LAG.) FEDDE., likewise (van de Laar 1975), we assume genetic material of all three closely related North American species to be present in invasive populations. We suppose that plant breeding has created genetic variability by multiple introductions and hybridisation. Consequently, a putative genetic bottleneck during introduction may have been overcome facilitating the invasion. Using microsatellite markers we want to look for hybridisation in the invasive European populations and for genetic variability within and between the invasive populations in contrast to native populations of all putative parent species of invasive *Mahonia*. No microsatellite markers have been developed for *Mahonia* up to now. Here we describe the development of ten microsatellite loci for invasive *Mahonia* and the applicability of these loci for three *Mahonia* species, and for *Berberis vulgaris* L., the only species that is native to Central Europe and closely related to *Mahonia* (subtribe Berberidinae).

2.3   Material and Methods

Development of microsatellite loci was carried out by ECOGENICS GmbH (Zürich, Switzerland) using fresh leaf material from one single cultivated *Mahonia aquifolium* cultivar. Genomic DNA was extracted by a modified CTAB method. After CTAB incubation the DNA was precipitated with potassium-acetate. The supernatant was extracted with chloroform/ isoamyl alcohol and an ethanol precipitation was performed. An enriched library was made from size selected genomic DNA ligated into SAULA/SAULB-linker (5'-GCGGTACCCGGGAAGCTTGG/ 5'- GATCCCAAGCTT CCCGGGTACCGC (Armour et al. 1994) and enriched by magnetic bead selection with biotin-labelled $(CA)_{13}$ and $(GA)_{13}$ oligonucleotide repeats (Gautschi et al. 2000a; Gautschi et al. 2000b). Of 576 recombinant colonies screened, 118 gave a positive signal after hybridisation. Plasmids from 72 positive clones were sequenced and primers were designed for 22 microsatellite inserts from which 11 primers were selected after an initial screening with 5 invasive *Mahonia* samples from central Germany. Primers were tested for polymorphism on 20 individuals of one invasive *Mahonia* population from north-east Germany. DNA was extracted using the Plant DNA extraction mini kit (QIAGEN). Polymerase chain reaction (PCR) was performed in 96-well plates using a Primus 96 plus (MWG BIOTECH) or a Mastercycler gradient (EPPENDORF) Termocycler. 10μL reactions contained 1-10ng genomic DNA, 1 pmole of each forward and reverse primer (MWG), 4 μL Multiplex PCR Kit (QIAGEN) and 3 μL H_2O. Fragments were separated on an ABI 310 genetic analyser (APPLIED BIOSYSTEMS)

with the size standard GeneScan 500 ROX (APPLIED BIOSYSTEMS). For visuali-
sation of fragments each forward primer was fluorescent labelled with FAM (Mahonia
CA03, CA18, CA30, GA04), JOE (CA40, GA05, GA31, GA33) or TAMRA (CA22,
CA43, GA36). Optimal annealing temperature (T_a) for each primer was ascertained
using a gradient from 55°C to 65°C. I grouped together primers with the same T_a to
multiplex reactions leading to three primer combinations: 1. GA31 and CA30 $(T_a =
57°C)$; 2. CA40, GA04, CA18 and CA22 $(T_a = 60°C)$; 3. GA33, GA05, CA03, GA36
and CA43 $(T_a = 63°C)$. In these reactions we reduced the amount for some primers
(CA18, GA05, GA31) to 0.5 pmole because of unequal amounts of amplification
products within multiplex reaction. The PCR-program was the following: 95°C for 15
minutes, 30 cycles of 94°C for 30 s, T_a for 90 s, 72°C for 60 s and at the end 60°C for
30 minutes. For statistical analyses locus CA22, which gave irreproducible results, was
excluded. There were ten primers left, which gave interpretable PCR products (Table 1).

2.4   Results and Discussion

No homozygous null allele was observed. Expected and observed heterozygosity was
calculated using MSA software (Dieringer and Schlötterer 2002) and inbreeding
coefficient F_{IS} was calculated using FSTAT 2.9.3.2. software (Goudet 1995). Allele
numbers, which ranged between 3 and 13 alleles per locus, indicated a high genetic
diversity in the invasive population, which was mirrored in equally high values of
observed (mean $H_o = 0.573$) and expected heterozygosities (mean $H_e = 0.620$). Only
two loci (CA03, CA30) showed a significant F_{IS} value, which may be due to non
random sampling or null alleles but probably not to inbreeding, since $M.$ $aquifolium$ is
supposed to be an outbreeding species like the whole genus (Burd 1994). The test of
linkage disequilibrium was performed using GENEPOP software, Version 2 (Raymond
and Rousset 1995). All 45 combinations of loci were tested, but only one combination
of loci (GA31 and CA40) was significantly linked. In order to check the applicability of
the microsatellite markers for the putative parent species of invasive $Mahonia$ and the
related genus $Berberis$, we tested the markers in $M.$ $aquifolium,$ $M.$ $repens$ and $M.$
$pinnata$ from North America (eight individuals from one population, each), and at nine
individuals from five European populations of $Berberis$ $vulgaris$. The results (Table 2)

showed that the loci seem to be conserved within *Mahonia* and partly also in *Berberis*. However, homozygous null alleles were found in two, one and three loci, respectively in *M. aquifolium, M. repens* and *B. vulgaris*. Three markers (CA18, CA40, GA36) did not reveal any fragments in *B. vulgaris*. We will use the microsatellite markers to analyse the genetic variability in invasive and native *Mahonia* populations and to look for a hybrid-origin of invasive populations.

Table 1: Microsatellite loci from invasive *Mahonia* based on one population of 20 individuals from north-east Germany

Locus	Repeat motif in sequenced clone	Primer sequence (5'-3')	No. of alleles	Size range (bp)	H_o	H_e	F_{IS}	P-value	EMBL accession number
Mahonia CA03	$(AT)_3(GT)_{18}$	F: GGGGTGTGACCGTTTTTATG R: CAATGCCCGAAAGTTACGTC	7	130-176	0.632	0.838	0.251	0.0241	AM233740
Mahonia CA18	$(TA)_3(TG)_{27}$	F: TCAATTCTTTGAGTTAGGGTTTTG R: CCAATGACGTTAAATCCATACG	9	162-204	0.750	0.849	0.119	0.1723	AM233741
Mahonia CA30	$(CA)_{31}$	F: TGCATTTTCGACCCATCTAC R: TCTCCTCACATGCAACAAAAG	13	92-152	0.55	0.908	0.400	0.0025	AM233742
Mahonia CA40	$(CA)_{23}$	F: CGTATCTTTACTGTGAAATGGTGAG R: AGGTTAAATAAATTTCATCAATCACTC	9	113-135	0.750	0.804	0.069	0.3478	AM233743
Mahonia CA43	$(CA)_{12}$	F: TCCGCTTTCCACTTACCATC R: GGATGAGGGAGGTGTAACAATG	5	112-128	0.421	0.451	0.068	0.4313	AM233744
Mahonia GA04	$(GA)_{18}$	F: ACCCATTGGAGCTCTCTCAG R: TTGATTTTGAAGCCGAGATG	8	106-128	0.75	0.806	0.072	0.3475	AM233745
Mahonia GA05	$(CA)_{15}$	F: AGTCATCCCTCCATCATTCG R: TGTGAGAGCTCTGTTGGACTG	3	141-151	0.632	0.482	-0.321	1.0000	AM233746
Mahonia GA31	$(GT)_{12}$	F: TCACAATAGTTTATTTGAGTTTATTTG R: CACTGTCTGGCTCAATTTTGTC	3	158-164	0.500	0.645	0.229	0.1278	AM233747
Mahonia GA33	$(CA)_{14}$	F: GATCAGGTCCATAATATCAAAGTTC R: CAGACAAGGAGAGTGCTTGTACC	4	217-225	0.316	0.324	0.027	0.4855	AM233748
Mahonia GA36	$(GT)_{16}$	F: ACGAGGGCTATACAGGAACC R: CCAAGTATGTCCAGTACCTCCAG	3	176-186	0.579	0.568	-0.021	0.6361	AM233749

F, forward primer; R, reverse primer; H_o, observed heterozygosity; H_e, expected heterozygosity; F_{IS}, inbreeding coefficient; P-value, probability that F_{IS} is not different from zero.

Isolation and characterisation of microsatellite markers

Table 1: Microsatellite loci from invasive *Mahonia* based on one population of 20 individuals from north-east Germany

Locus	Repeat motif in sequenced clone	Primer sequence (5'-3')	No. of alleles	Size range (bp)	H_o	H_e	F_{IS}	P-value	EMBL accession number
Mahonia CA03	$(AT)_3(GT)_{18}$	F: GGGGTGTGACCGTTTTTATG R: CAATGCCCGAAAGTTACGTC	7	130-176	0.632	0.838	0.251	0.0241	AM233740
Mahonia CA18	$(TA)_3(TG)_{27}$	F: TCAATTCTTTTGAGTTAGGGTTTTG R: CCAATGACGTTAAATCCATACG	9	162-204	0.750	0.849	0.119	0.1723	AM233741
Mahonia CA30	$(CA)_{31}$	F: TGCATTTTCGACCCATCTAC R: TCTCCTCACATGCAACAAAAG	13	92-152	0.55	0.908	0.400	0.0025	AM233742
Mahonia CA40	$(CA)_{23}$	F: CGTATCTTTACTGTGAAATGGTGAG R: AGGTTAAATAAATTTCATCAATCACTC	9	113-135	0.750	0.804	0.069	0.3478	AM233743
Mahonia CA43	$(CA)_{12}$	F: TCCGCTTTCCACTTACCATC R: GGATGAGGGAGGTGTAACAATG	5	112-128	0.421	0.451	0.068	0.4313	AM233744
Mahonia GA04	$(GA)_{18}$	F: ACCCATTGGAGCTCTCTCAG R: TTGATTTTGAAGCCGAGATG	8	106-128	0.75	0.806	0.072	0.3475	AM233745
Mahonia GA05	$(CA)_{15}$	F: AGTCATCCCTCCATCATTCG R: TGTGAGAGCTCTGTTGGACTG	3	141-151	0.632	0.482	-0.321	1.0000	AM233746
Mahonia GA31	$(GT)_{12}$	F: TCACAATAGTTTATTTGAGTTTATTTG R: CACTGTCTGGCTCAATTTTGTC	3	158-164	0.500	0.645	0.229	0.1278	AM233747
Mahonia GA33	$(CA)_{14}$	F: GATCAGGTCCATAATATCAAAGTTC R: CAGACAAGGAGAGTGCTTGTACC	4	217-225	0.316	0.324	0.027	0.4855	AM233748
Mahonia GA36	$(GT)_{16}$	F: ACGAGGGCTATACAGGAACC R: CCAAGTATGTCCAGTACCTCCAG	3	176-186	0.579	0.568	-0.021	0.6361	AM233749

F, forward primer; R, reverse primer; H_o, observed heterozygosity; H_e, expected heterozygosity; F_{IS}, inbreeding coefficient; P-value, probability that F_{IS} is not different from zero.

References

Ahrendt, L.W.A. (1961) *Berberis* and *Mahonia*. A taxonomic revision. *Journal of the Linnean Society of London, Botany,* **57**: 1-410.

Armour, J.A., Neumann, R., Gobert, S. and Jeffreys, A.J. (1994) Isolation of human simple repeat loci by hybridization selection. *Human Molecular Genetics,* **3**: 599-565.

Burd, M. (1994) Bateman principle and plant reproduction - the role of pollen limitation in fruit and seed set. *Botanical Review,* **60**: 83-139.

Dieringer, D. and Schlötterer, C. (2002) Microsatellite analyser (MSA): a platform independent analysis tool for large microsatellite data sets. *Molecular Ecology Notes,* **3**: 167-169.

Gautschi, B., Tenzer, I., Muller, J.P. and Schmid, B. (2000a) Isolation and characterization of microsatellite loci in the bearded vulture (*Gypaetus barbatus*) and cross-amplification in three Old World vulture species. *Molecular Ecology,* **9**: 2193-2195.

Gautschi, B., Widmer, A. and Koella, J. (2000b) Isolation and characterization of microsatellite loci in the dice snake (*Natrix tessellata*). *Molecular Ecology,* **9**: 2191-2193.

Goudet, J. (1995) FSTAT (Version 1.2): A computer program to calculate *F*- statistics. *Journal of Heredity,* **86**: 485-486.

Kowarik, I. (1992) Einführung und Ausbreitung nichteinheimischer Gehölzarten in Berlin und Brandenburg. *Verhandlungen Botanischer Vereine Berlin Brandenburg,* **3**: 1-188.

Raymond, M. and Rousset, F. (1995) Genepop (Version-1.2) - Population-Genetics Software for Exact Tests and Ecumenicism. *Journal of Heredity,* **86**: 248-249.

van de Laar, H.J. (1975) *Mahonia en Mahoberberis. Dendroflora,* **11/12**: 19-33.

Chapter 3

Genetic relationships among three native North-American *Mahonia* species, invasive *Mahonia* populations from Europe, and commercial cultivars

In cooperation with Harald Auge and Walter Durka

3.1   Abstract

Horticulture is one of the most important pathways for plant invasion. We used microsatellite markers to reveal the impact of plant breeding on *Mahonia aquifolium*, an invasive ornamental shrub. Since it was bred by hybridisation with the related species *M. repens* and *M. pinnata*, we compared populations of the three native species, various commercial cultivars and invasive populations. Invasive populations and cultivars were genetically differentiated from the native groups, but differences did not result from genetic bottlenecks. In cultivars but not in invasive populations, we proved genes from *M. pinnata*. No significant amount of *M. repens* genes were found in cultivars and invasive populations, but this result has to be viewed with caution because of the close relationship between native *M. aquifolium* and *M. repens*. We conclude that the evolution of invasive *Mahonia* populations was a result of restriction of gene pool during introduction, secondary release, and artificial selection, in combination with an increase of genetic diversity by plant breeders and by extensive gene flow.

3.2   Introduction

A major reason for changes of our floras is the spread of exotic plant species that were introduced intentionally as horticultural and agricultural plants (Preston et al. 2002; Mack 2000). In Germany cultivated plants make up 50 % of all neophytes and 70 % of those neophytes, which are established in natural habitats (Klotz et al. 2002). Two factors may particularly contribute to the success of cultivated plants in the invasion of new habitats. The first one is a mass effect of cultivated plants that are planted in very large numbers at various locations and are often protected by man from detrimental environmental effects. This causes a high propagule pressure at numerous sites and a high probability of invasion (Kowarik 2005; Mack 2000). Propagule pressure has generally been shown to be one of the few factors that can be identified to determine invasion success (Rejmanek 2000). Second, cultivation of plants does usually include evolutionary changes. Evolutionary changes are suggested to play a major role in plant invasion (Ellstrand and Schierenbeck 2000). Indeed, it has been repeatedly shown that invasive populations differ genetically from their ancestral populations in natural habitats (reviewed in Bossdorf et al. 2005).

Genetic differences between invasive and native populations may result from genetic bottlenecks after introduction (Barrett and Richardson 1986). This may result in reduced genetic diversity in the founder populations and a lower probability of persistence of the new invader (Allendorf and Lundquist 2003). In contrast to unintentionally introduced plant species, cultivated plants are not introduced randomly but selective. Wild individuals are selected because of their preferred phenotype which may lead to a limited but above-average fit subsample of genotypes being introduced in a new area. In cultivation, plants are intensively selected and modified by man resulting in further changes of their genetic makeup. Artificial selection for fitness-related traits such as flower size, seed number or cold tolerance, may enhance species success not only in gardens but also in natural habitats (e.g. Kitajima et al. 2006). In addition, a common method in plant breeding is interspecific hybridisation, which results in an increased genetic variability and novel genotypes that are potentially better adapted to the new environment (Ellstrand and Schierenbeck 2000). Thus, invasions may be facilitated by hybridisation because of a few well adapted genotypes and/or because hybrid populations overcome genetic bottlenecks and are thus able to respond to changing environmental conditions. Several studies have shown that hybrids, representing new genetic entities, may colonise territories were the parent species do not occur (e.g. Milne and Abbott 2000; Neuffer et al. 1999; Hollingsworth et al. 1998).

Although cultivated plants are above average successful in invasion, only a small proportion of all cultivated species is likely to spread (Kowarik 2005). A proscription of all cultivated plants would be needless and inappropriate. However, this small number of invasive species may cause an ecological and economic impact

(Kowarik 2005) and it is important to understand how cultivation of ornamentals facilitates plant invasion. Studies of the evolution of invasive species should contribute to our understanding of invasiveness (Ellstrand and Schierenbeck 2000). However, the role of plant breeding in invasion success has rarely been studied (but see Kitajima et al. 2006), although some of the most serious invaders are ornamentals. The fast spread of *Impatiens glandulifera* is likely to result from dispersal of garden plants by man (Perrins et al. 1993) and *Rhododendron ponticum* was hybridised by breeders with cold tolerant related species and that may have facilitated its invasion in Great Britain (Milne and Abbott 2000).

In our study we investigated the woody plant, *Mahonia aquifolium* Pursh. (Nutt.) (Berberidaceae). It was introduced from North America to Europe as an ornamental because of its evergreen leaves, yellow flowers and blue berries and is one of the most successful alien shrubs in central and eastern Germany today (Kowarik 1992). In cultivation, the related North American species, *M. repens* and *M pinnata*, were hybridised with *M. aquifolium*. Different cultivars with various characteristics in flowering, clonal growth and resistance against parasites arose (Houtman et al. 2004) and were frequently planted in gardens, parks and along roads. *M. aquifolium* produces many fleshy fruits which are eaten by birds that disperse seeds also into adjacent habitats. Today, the species is spreading and invades anthropogenic and natural habitats (Kowarik 1992), and propagates not only by seeds but also by stolons and stem layering (Auge and Brandl 1997). This is less known from native *M. aquifolium* but commonly from *M. repens* (Ahrendt 1961) and also found in some cultivars (Houtman et al. 2004). It is assumed that invasive populations descend from cultivars, which are supposed to be hybrids (van de Laar 1975; Ahrendt 1961). Therefore it is likely that invasive populations consist of hybrids, although the invasive shrubs are referred to as *M. aquifolium*. Thus, *Mahonia* is a well-suited case study for investigating the role of plant breeding for invasion success.

In this study we explore whether invasion success of *Mahonia* populations is a result of evolution by plant breeding. We investigated the following questions. (1) Was there a genetic bottleneck after introduction of *Mahonia* to Europe? (2) Is there genetic differentiation of invasive *Mahonia* populations from the native species, and how do cultivated individuals arrange to native species and invasives? (3) Do the invasive populations consist of hybrids between the three *Mahonia* species?

3.3   Material and Methods

3.3.1   Species

The genus *Mahonia* NUTT. (Berberidaceae) comprises fleshy-fruited evergreen shrubs with pinnate leaves. The genus is treated distinct from the genus *Berberis* (Ahrendt 1961), but inclusion into *Berberis* is also common (Kim et al. 2004; Laferriere 1997). Oregon Grape, *Mahonia aquifolium* (PURSH) NUTT., is native to western North America (Figure 1), has a simple and erect stem that reaches 1.80 m in height with leaves that are shiny above and dull underneath (Ahrendt 1961; Piper 1922). *M. aquifolium* was introduced into Europe for ornamental purposes in 1822 (Hayne 1822, cited in Kowarik 1992) and repeatedly later on. The first spontaneous occurrence outside gardens was observed in 1860 after a time lag of 38 years (Kowarik 1992). The species was extensively hybridised by plant breeders with related, North American species, in particular with *M. repens* (LINDL.) G.DON (Ahrendt 1961) and *Mahonia pinnata* (LAG.) FEDDE., as indicated by many cultivated hybrids (van de Laar 1975). *M. repens* is morphologically very similar to *M. aquifolium* and some specimens are difficult to assign to one of the two species (Ahrendt 1961), *M. repens* reaches only 90 cm in height and grows usually more stoloniferous than *M. aquifolium* (Ahrendt 1961). The leaves are mostly dull above (Piper 1922). *M. pinnata* reaches 3 m height (Ahrendt 1961) and has shiny leaves above and underneath. In contrast to the other two species the first leaflets of the pinnate leaf arise near base of petiole (Munz 1959). The breeding and hybridisation of the three *Mahonia* species resulted in many cultivars (van de Laar 1975) of the three species and their hybrids (*M. x decumbens = M. aquifolium* x *M. repens*; *M. x wagneri = M. aquifolium* x *M. pinnata*).

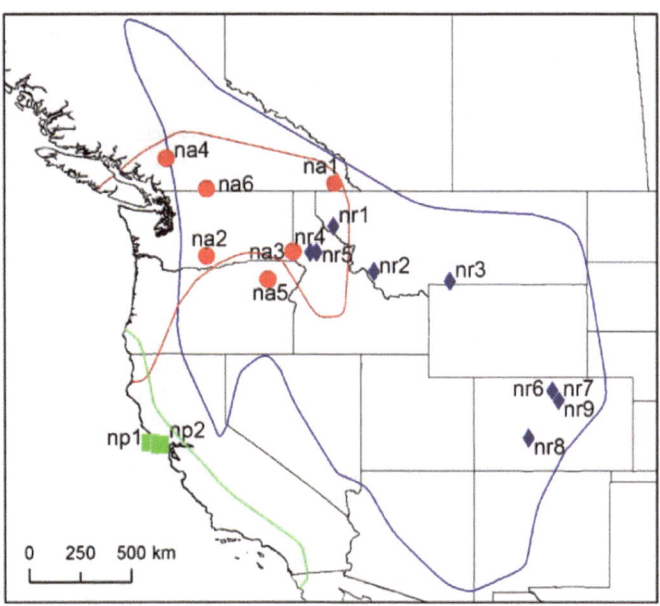

Figure 1: Map of the distribution areas of the three native species after Whittemore (1997) and the locations were populations were sampled. red ● *M. aquifolium*, blue ◆ *M. repens*, green ■ *M. pinnata*,

3.3.2   Sampling and genetic analyses

We analysed *Mahonia* individuals of five taxa including three native species, invasive individuals and cultivars (Table 1 and Table 2). In the native range in North America, individuals of *M. aquifolium*, *M. repens* and *M. pinnata* were sampled in six (n = 65 samples), nine (n = 119) and two (n = 34) populations, respectively (Figure 1). Invasive *Mahonia* were sampled from 23 invasive populations (n = 416 individuals) in Germany and the Czech Republic. Within populations individuals were sampled randomly and attention was paid to sample spatially separated plant individuals. However, in some dense populations, individuals were not clearly separated.

Table 1: Sampled populations of invasive *Mahonia* and native *M. aquifolium, M. repens* and *M. pinnata*. The first letter of the site code identifies the origin (i = invasive; n = native), the second letter indicates the species, a = *M. aquifolium*, p = *M. pinnata* and r = *M. repens*.

code	origin	species	site	location	n
i1	invasive		Germany; Barby	51.6 N; 11.6 E	20
i2	invasive		Germany; Bocka-Neustaedtel	51.1 N; 14.1 E	22
i3	invasive		Germany; Berlin	52.3 N; 13.2 E	19
i4	invasive		Germany; Buckow	52.3 N; 14.1 E	20
i5	invasive		Germany; Drebkau	51.4 N; 14.1 E	20
i6	invasive		Germany; Drebkau1	51.4 N; 14.1 E	8
i7	invasive		Germany; Duebener Heide	51.4 N; 12.3 E	20
i8	invasive		Germany; Duisburg	51.2 N; 06.5 E	13
i9	invasive		Germany; Herzfelde	52.3 N; 13.5 E	10
i10	invasive		Germany; Hitzhausen	52.2 N; 08.2 E	9
i11	invasive		Germany; Hornburg	52.0 N; 10.4 E	16
i12	invasive		Germany; Jena	50.6 N; 11.4 E	20
i13	invasive		Germany; Kirchbrak	51.6 N; 09.4 E	19
i14	invasive		Germany; Halle-Lieskau	51.3 N; 11.6 E	10
i15	invasive		Germany; Liepe	52.5 N; 13.6 E	22
i16	invasive		Germany; Linz am Rhein	50.3 N; 07.2 E	28
i17	invasive		Germany; Lueneburg	53.1 N; 10.2 E	22
i18	invasive		Germany; Mannheim	49.3 N; 08.3 E	22
i19	invasive		Germany; Neuhaus a.d. Pegnitz	49.4 N; 11.3 E	26
i20	invasive		Czech Republik; Prag	50.7 N; 14.3 E	9
i21	invasive		Germany; Rothenburg	51.4 N; 11.5 E	17
i22	invasive		Germany; Suckow	53.3 N; 12.2 E	24
i23	invasive		Germany; Zierenberg	51.2 N; 09.2 E	20
na1	native	*M. aquifolium*	British Columbia; Tie Lake	49.3 N; 115.2 W	18
na2	native	*M. aquifolium*	Washington; Cle Elum	46.2 N; 120.8 W	15
na3	native	*M. aquifolium*	Idaho; Harvard	46.4 N; 117.0 W	7
na4	native	*M. aquifolium*	Oregon; Viento	50.4 N; 122.6 W	9
na5	native	*M. aquifolium*	Oregon; LaGrande	45.2 N; 118.1 W	6
na6	native	*M. aquifolium*	British Columbia; Manning Park	49.1 N; 120.8 W	10
np1	native	*M. pinnata*	California; Bodega Bay	38.2 N; 123.3 W	13
np2	native	*M. pinnata*	California; Tomales Bay	38.1 N; 122.9 W	21
nr1	native	*M. repens*	Montana; Bear Lake	47.6 N; 115.3 W	7
nr2	native	*M. repens*	Montana ; Blackfoot River	45.6 N; 113.4 W	15
nr3	native	*M. repens*	Montana; Boulder River	45.2 N; 110.1 W	14
nr4	native	*M. repens*	Idaho; Deary	46.5 N; 116.3 W	10
nr5	native	*M. repens*	Idaho; Deer Road	46.5 N; 116.0 W	9
nr6	native	*M. repens*	Colorado; Poudre Canyon	40.4 N; 105.5 W	21
nr7	native	*M. repens*	Colorado; Big South Trailhead	40.4 N; 105.5 W	19
nr8	native	*M. repens*	Colorado; Crested Butte Mountain	38.5 N; 106.5 W	14
nr9	native	*M. repens*	Colorado; Middle Saint Vrain Valley	40.1 N; 105.3 W	10

Table 2: Cultivars included in the study. Selection year was taken from Houtman et al. (2004). Information about selection of c19-c21 was given by breeder himself.

code	species	cultivar	year of selection	n
c1	*M. aquifolium*			11
c2	*M. aquifolium*	Apollo	1973	13
c3	*M. aquifolium*	Atropurpurea	1915	11
c4	*M. aquifolium*	Darthil ®	2000	1
c5	*M. aquifolium*	Dart's Distinction	1970	1
c6	*M. aquifolium*	Dart's Quickstep	1987	1
c7	*M. aquifolium*	Euro	1996	1
c8	*M. aquifolium*	Golden Pride	Unknown	1
c9	*M. aquifolium*	Green Ripple	1970	3
c10	*M. aquifolium*	Hastings Elegant	Unknown	3
c11	*M. aquifolium*	Hans-Karl Möhring	1984	4
c12	*M. aquifolium*	Juglandifolium	Unknown	2
c13	*M. aquifolium*	Jupiter	1978	3
c14	*M. aquifolium*	Maqu	1970	5
c15	*M. aquifolium*	Marijke	1993	1
c16	*M. aquifolium*	Mirena	1979	10
c17	*M. aquifolium*	Orange Flame	1965	2
c18	*M. aquifolium*	Smaragd	1978	9
c19	*M. aquifolium*	Typ1	1999	3
c20	*M. aquifolium*	Typ2	1999	3
c21	*M. aquifolium*	Typ3	1999	3
c22	*M. aquifolium*	Undulata	1930	7
c23	*M. aquifolium*	Versicolor	Unknown	3
c24	*M. x decumbens*	Bokrafood ®	2001	1
c25	*M. x decumbens*	Bokrahawk ®	Unknown	1
c26	*M. x decumbens*	Bokrarond ®	2005	1
c27	*M. x decumbens*	Bokrasio ®	2003	2
c28	*M. x decumbens*	Cosmo Crawl	1992	1
c29	*M. x decumbens*	Nr17	Unknown	1
c30	*M. x decumbens*	Pixie	1994	1
c31	*M. x hybrida*	Hybrida	cultural bastard	2
c32	*M. pinnata*	Ken Howard	Unknown	3
c33	*M. repens*		Unknown	4
c34	*M. x wagneri*	Darts Flashlight	1993	1
c35	*M. x wagneri*	Fireflame	1965	1
c36	*M. x wagneri*	Moseri	1895	1
c37	*M. x wagneri*	Pinnacle	1930	4
c38	*M. x wagneri*	Sunset	1998	4
c39	*M. x wagneri*	Vicaryi	1931	1

Furthermore we sampled 127 individuals that belonged to 39 different cultivars from Botanical Gardens and commercial nurseries (Table 2). We either sampled leaves directly in the field which were dried and stored in silica gel (20 populations and cultivars) or we sampled seeds (20 populations). In the latter case seedlings were grown from cold stratified seeds in a climate chamber with a 14 h/10 h day/night cycle at 15°C/10°C. When the seedlings had secondary leaves we harvested and stored them at -80°C. From each mother plant only one seedling was analysed. DNA was extracted from dried or frozen leaves with the Plant DNA extraction mini kit (QIAGEN). A total of 761 individuals were genotyped at eight microsatellite loci (CA03, CA18, CA40, CA43, GA05, GA31, GA33, GA36) as described previously (Roß and Durka 2006).

3.3.3   Data analysis

We measured a number of genetic parameters to compare genetic diversity between invasive and native populations. We analysed the number of alleles and observed and expected heterozygosity using MSA 3.0 (Dieringer and Schlötterer 2002). Allelic richness (A_r), a measure of allelic diversity corrected for sample size, was calculated with FSTAT 2.9.3.2. software (Goudet 1995). Departure from Hardy-Weinberg-equilibrium was also tested with FSTAT. Samples with identical multilocus genotypes were regarded as clones. Genetic parameters at population level were calculated using each multilocus genotype once. Differences of genetic parameters between invasive and native populations and between the species nested within status (invasive or native) were calculated by a nested ANOVA using the GLM procedure in SAS 9.1. (SAS Institute, Cary, NC, USA).

We calculated the number of private alleles at species level. Private alleles were defined as alleles present in more than one population of a species and in no other species. Overall genetic differentiation among populations was assessed with F-statistics (Weir and Cockerham 1984; Wright 1951) using FSTAT. Genetic differentiation between populations within species was estimated as the pair wise F_{ST}-value and compared between the taxa using FSTAT with 1000 permutations. We excluded *M. pinnata* populations from this analysis due to low sample size. We tested for isolation by distance for the three native species with a Mantel test with 2000 randomisations using FSTAT. We also assessed genetic differentiation among species and invasives by a hierarchical analysis of molecular variance (AMOVA) with Arlequin 2.0 (Schneider et al. 2000) with populations nested in species.

We used a Principle Components Analysis (PCA) and a model-based clustering method to analyse the relationship among native taxa, invasives and cultivars. PCA was carried out using PCAGEN 1.2 (Goudet 1999). Furthermore, we clustered all individuals using STRUCTURE 2.0 (Pritchard et al. 2000). This software uses a model-based Bayesian procedure to assign individuals into K clusters based on their multilocus genotypes. To identify the most probable number of clusters the algorithm was run with values of K from 1 to 14 ten times each. We used the admixture model with a length of burnin period of 10,000 and 10,000 iterations and the prior information about the populations. The posterior probabilities of K, (L(K)) and ΔK calculated according to Evanno (2005) were used as indicators of the most probable K value. The whole data set including natives, invasive populations and cultivars was analysed and visualized using the DISTRUCT program (Rosenberg 2004).

3.4   Results

3.4.1   Genetic variation

At eight microsatellite loci we detected 187 different alleles in a total of 761 individuals. The number of alleles per locus ranged between ten (GA05) and 37 (CA03 and CA18). At species level we found 131 alleles in *M. aquifolium*, 144 alleles in *M. repens* and 69 alleles in *M. pinnata*. In European samples a smaller number of alleles was found with 101 alleles in invasives and 106 alleles in cultivars. The frequency of species specific alleles was low with 6 (5.1 %) private alleles in *M. aquifolium* and 16 (11.6%) private alleles in *M. repens*, but no allele was specific to *M. pinnata*. However, several alleles were common in one species and rare in the others, or species were characterised by the absence of an otherwise common allele. In seven out of 40 populations (three invasive populations, two *M. repens* populations and two *M. pinnata* populations) several samples shared the same multilocus genotype indicating clonal propagation. Populations were highly diverse with mean H_e = 0.60 ± 0.06 and 0.65 ± 0.02 and mean H_o 0.48 ± 0.04 and 0.57 ± 0.02 (means ± s.e.) in the native and invasive populations, respectively (Table 3). Most F_{IS}-values were significant, with mean F_{IS} = 0.17 and 0.12 in native and invasive taxa, indicating slight departure from Hardy-Weinberg expectations, which may be due to null-alleles (Roß and Durka 2006) but not due to inbreeding, because self pollination in *Mahonia* does rarely result in fruit production (Monzingo 1987, H. Auge unpublished data). Native and invasive taxa did not differ significantly in number of alleles per locus, allelic richness, expected and observed heterozygosity and inbreeding coefficient. Furthermore, there were no significant differences between the native species in these traits (ANOVA, A: $p = 0.374$, Ar: $p = 0.059$, F_{IS}: $p = 0.725$) except in H_e ($p = 0.037$) and H_o ($p = 0.029$) with lowest values in *M. pinnata* and highest values in *M. aquifolium* (Table 3).

Table 3: Genetic diversity at eight microsatellite loci of invasive and native *Mahonia* populations: Sample size (N), number of multilocus genotypes (N_{GT}) Observed heterozygosity (H_o), expected heterozygosity (H_e), number of alleles per locus (A), allelic richness (A_r) based on five individual and inbreeding coefficient (F_{IS}, *: $p < 0.05$, **: $p < 0.01$, ***: $p < 0.001$). Genetic parameters were calculated on multilocus genotypes instead of individuals. Mean values of native species and invasive populations are least square means calculated by ANOVA.

population	N	N_{GT}	H_o	H_e	A	A_r	F_{is}
invasive							
i1	20	20	0.59	0.67	5.3	3.8	0.119**
i2	22	16	0.63	0.52	4.0	2.8	-0.214
i3	19	19	0.55	0.67	5.4	3.7	0.183***
i4	20	20	0.58	0.62	5.5	3.7	0.063
i5	20	20	0.50	0.64	4.8	3.5	0.210***
i6	8	8	0.53	0.60	3.9	3.4	0.158*
i7	20	20	0.52	0.64	5.0	3.5	0.195***
i8	13	13	0.56	0.67	6.0	4.0	0.171**
i9	10	10	0.50	0.62	4.1	3.3	0.191**
i10	9	9	0.50	0.64	4.4	3.7	0.213**
i11	16	16	0.65	0.72	5.9	4.1	0.092*
i12	20	20	0.61	0.65	5.8	3.9	0.061
i13	19	19	0.64	0.68	5.6	4.1	0.066
i14	10	10	0.55	0.68	5.5	4.2	0.198**
i15	22	22	0.64	0.70	6.4	4.1	0.087*
i16	28	28	0.52	0.66	6.8	4.0	0.218***
i17	22	20	0.69	0.71	6.3	4.1	0.031
i18	22	22	0.59	0.72	7.4	4.4	0.176***
i19	26	26	0.62	0.65	5.9	3.7	0.053
i20	9	9	0.57	0.66	4.9	4.0	0.137*
i21	17	17	0.53	0.64	4.6	3.5	0.182***
i22	24	14	0.54	0.54	4.1	3.1	0.011
i23	20	20	0.58	0.67	6.0	4.0	0.130**
mean invasive			**0.57**	**0.65**	**5.4**	**3.8**	**0.119**

Continuance **Table 3:**

population	N	N_{GT}	H_o	H_e	A	A_r	F_{is}
native							
M. aquifolium							
na1	18	18	0.39	0.73	7.6	4.7	0.362***
na2	15	15	0.52	0.75	7.1	4.7	0.234***
na3	7	7	0.61	0.58	4.5	3.9	-0.074
na4	6	6	0.48	0.77	5.0	4.7	0.380***
na5	9	9	0.56	0.71	5.8	4.6	0.173**
na6	10	10	0.55	0.67	4.4	3.6	0.100
mean			*0.56*	*0.70*	*5.7*	*4.4*	*0.200*
M. pinnata							
np1	13	7	0.49	0.39	2.6	2.3	0.200*
np2	21	16	0.53	0.66	6.5	4.2	0.195***
mean			*0.42*	*0.52*	*4.6*	*3.2*	*0.198*
M. repens							
nr1	7	7	0.47	0.76	4.8	4.3	0.308***
nr2	15	15	0.40	0.50	5.5	3.6	0.103*
nr3	14	14	0.43	0.66	5.5	3.9	0.306***
nr4	10	10	0.65	0.78	6.6	4.9	0.214***
nr5	9	9	0.35	0.74	6.3	4.9	0.282***
nr6	21	4	0.48	0.30	1.9	1.7	-0.697
nr7	19	19	0.45	0.62	7.0	4.1	0.253***
nr8	14	6	0.38	0.29	2.8	2.0	0.081
nr9	10	10	0.38	0.54	3.5	3.0	0.281***
mean			*0.47*	*0.58*	*4.9*	*3.6*	*0.126*
mean native			**0.48**	**0.60**	**5.1**	**3.7**	**0.173**

As expected from the large proportion of shared alleles, the species where significantly but only weakly differentiated with 10.3 % of genetic variation residing among species (AMOVA: $\Phi_{CT} = 0.103$) and 12.3 % of variation residing among populations within species ($\Phi_{SC} = 0.137$) (Table 4). Populations were weakly but significantly structured with overall F_{ST} values of 0.074 ± 0.006 for invasive populations and 0.093 ± 0.015 for native *M. aquifolium*. *M. repens* had an overall F_{ST} value of 0.329 ± 0.058 and, thus, was significantly ($p = 0.007$) more structured than the other taxa. There was no isolation by distance in invasive *Mahonia* ($p = 0.328$) and in *M. aquifolium* ($p = 0.666$), whereas *M. repens* showed a strong correlation of genetic and geographic distance ($r^2 = 0.287$, $p = 0.002$).

Table 4: Hierarchical analysis of molecular variance (AMOVA) for 600 individuals of three native species and invasive *Mahonia* populations. Variance components and explained variation between taxa, among populations within taxa and within populations.

Source of variation	d.f.	Sum of squares	Variance components	Percentage of variation
Among taxa	3	252.077	0.31707	10.28
Among populations within taxa	36	516.104	0.37794	12.26
Within populations	1228	2933.260	2.38865	77.46
Total	1267	3701.442	3.08365	

3.4.2   Relationship of native species, cultivars and invasives

The analysed taxa were not clearly separated by PCA (Figure 2). While native *M. aquifolium* was separated from native *M. pinnata* along the first axis (score mean ± standard deviation: *M. aquifolium* -0.10 ± 0.08; *M. pinnata* 1.03 ± 0.04), *M. repens* widely scattered along the first (0.80 ± 0.57) and second axis (0.35 ± 0.81). *M. repens* was separated into two groups along the second axis (0.99 ± 0.34 and -0.44 ± 0.28, respectively). These two groups of *M. repens* correspond to two areas that were sampled in the south and north of the species range (Figure 1). Cultivars were highly divers (first axis: 0.07 ± 0.39; second axis: 0.08 ± 0.08) with *M. aquifolium* located within them. Invasive *Mahonia* populations arranged mostly within cultivars, next to native *M. aquifolium* (fist axis: -0.31 ± 0.13; second axis: -0.17 ± 0.21).

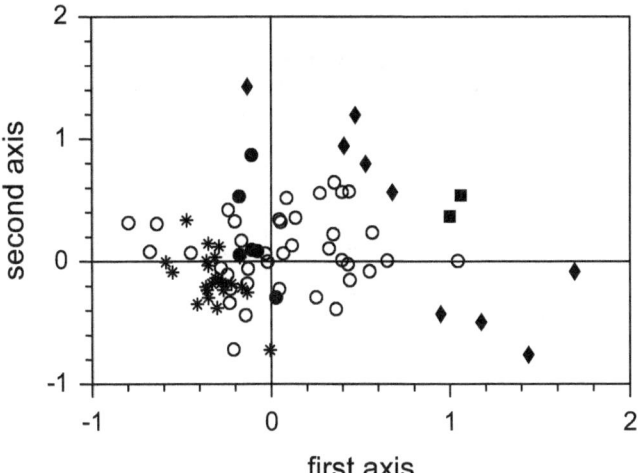

Figure 2: Principle components analysis (PCA) of allele frequencies at at eight microsatellite loci. ● *M. aquifolium*, ◆ *M. repens*, ■ *M. pinnata*, O cultivars, ✳ invasive populations. The first and second axis explained 19.26 % and 12.37 % of total variation. The analysed taxa were not clearly separated. Nevertheless, each native species grouped apart, *M. repens* was split in two groups. Cultivars were widely scattered with invasive *Mahonia* populations arranged mostly within cultivars

In the STRUCTURE analyses we found a similar pattern of grouping, even though the ΔK analysis revealed no definite number of groups. The log likelihood of K increased monotonously with increasing K from 1 to 14 (Figure 3). ΔK showed a peak at $K = 2$. However, the separation in only two groups did not allow to address the affiliation of invasive populations and cultivated plants to native species. Therefore, we plotted the results for $K = 2$ to $K = 6$ and, thus, zoomed into the genetic relationship of the analysed individuals from coarse ($K = 2$) to fine structure ($K = 6$) (Figure 4).

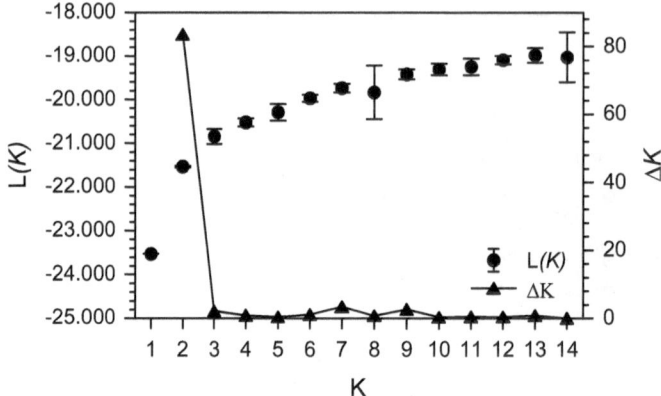

Figure 3: Graphical method to identify the true *K* (Evanno et al. 2005) from STRUCTURE analyses. Mean L(*K*) (± s.d.), the posterior probability of the data for a given *K* over 10 runs of each *K* (left axis), and Δ*K*, the standardised second order rate of change of L(*K*) (right axis). The log likelihood increased monotonously with increasing *K*. Δ*K* showed one peak at *K* = 2.

At *K* = 2, the coarse structure revealed two clusters which were built by native *M. repens* and *M. pinnata* on the one hand and invasive populations on the other hand. This separation was stronger than the separation of the native species. Native *M. aquifolium* showed admixture of both gene pools. Some cultivars clustered to *M. repens* and *M. pinnata*, but most cultivars clustered to the group of invasive *Mahonia* individuals. Within the group of native taxa, at all *K*-values > 2, four *M. repens* populations formed a consistent cluster. This strong splitting of native *M. repens* mirrored the geographical separation of the southern populations, as indicated by PCA analysis, before. The assignment of the *M. pinnata* populations was ambiguous. They clustered either with southern *M. repens* (*K* = 3) or with northern *M. repens* (*K* = 4) or formed an own gene pool (*K* = 5).

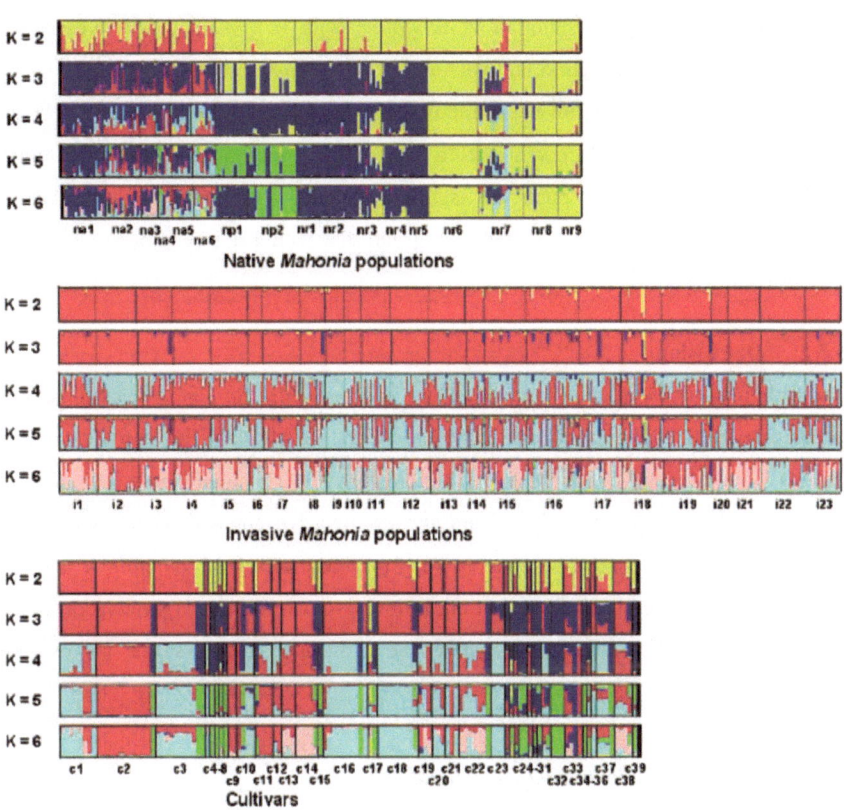

Figure 4: Estimated membership probability for 761 *Mahonia* genotypes for K genetic clusters identified by STRUCTURE analyses. Individuals are shown by vertical bars representing the proportional contribution of the K clusters to their genotype. Populations are separated by black lines. All individuals were introduced in all analyses. The three blocks show native populations, invasive populations and cultivars, respectively. The five graphs show five representative runs for $K = 2$ to $K = 6$.

At each K, native *M. aquifolium* did not represent an homogenous gene pool but showed admixture of northern *M. repens*, invasives' and cultivars' gene pools. The *M. aquifolium* population na1 showed stronger admixture of northern *M. repens* than other *M. aquifolium* populations. Most cultivars shared the group with invasive individuals. However, native *M. pinnata* gene pool was also found in cultivated individuals, namely in the *M pinnata* cultivar (c32), in *M. pinnata* hybrids (c34 – c39) as well as in *M. aquifolium* cultivars (c2, c3, c5 – c8, c15-c17) and in *M. repens* hybrids (c24 – c30).

This result indicated hybridisation of *M. aquifolium* with *M. pinnata* and the presence of *M. pinnata* gene pool in hybrid cultivars and *M. aquifolium* cultivars. *M. pinnata* gene pool was not detected in invasive populations. Furthermore, we found only small proportions of northern *M. repens* in cultivars and hardly any in invasive populations. Southern *M. repens* gene pool was neither detected in cultivars nor in invasive populations. Some cultivars assigned to a group that would not be expected by there species identification.

3.5   Discussion

The main results of our study were: (1) There is no evidence for a genetic bottleneck at population level after introduction of *Mahonia* to Europe. (2) The native *Mahonia* species have largely overlapping gene pools and were significantly but weakly differentiated. The majority of cultivars and the invasive populations formed a gene pool different from the native species. (3) Hybridisation of *M. aquifolium* and *M. pinnata* was displayed in cultivars but not in invasive populations. Hybridisation of *M. aquifolium* and *M. repens* could neither be proved in cultivars nor in invasive populations.

3.5.1   *Genetic diversity in invasive populations*

The differences between invasive and native *Mahonia* populations were not a result of a genetic bottleneck, because the genetic diversity was not significantly reduced in invasive populations. However, at the species level more alleles were found in natives than in cultivars and invasives. Thus, our results confirm the results of Bossdorf et al. (2005) that most invasions of plant species are not associated with an overall genetic bottleneck. Many invasions come about by multiple introductions that prevent a genetic bottleneck (Durka et al. 2005; Maron et al. 2004; Neuffer et al. 1999). In addition, inter- and intraspecific hybridisation can enhance genetic variation in invasive populations (Ellstrand and Schierenbeck 2000). We suppose that multiple introduction as well as hybridisation of *M. aquifolium*, *M. repens* and *M. pinnata* affected the genetic makeup of invasive *Mahonia* populations.

3.5.2   Relationships among native species

We presented evidence that North American native *Mahonia* species are differentiated at microsatellite loci (Figure 2 and Figure 4). *M. aquifolium* and *M. repens* were separated by private alleles and *M. pinnata* by the lack of certain alleles and by allele frequencies. The weak characterisation of *M. pinnata* by private alleles is likely owing to the low number of investigated populations. In the STRUCTURE analysis no definite *K* could be detected, indicating subtle continuous structure rather than distinct gene pools. The strongest division found was that between the northern and southern subranges of *M. repens* (Figure 4) which also corresponds to the high F_{ST}-value among *M. repens* populations compared to *M. aquifolium*. Whether there is indeed a clear cut geographical separation of distinct gene pools within *M. repens* or rather a clinal pattern, as indicated by the isolation by distance relationship, remains an open question. However, these findings are consistent with pronounced morphological variability among *M. repens* from different parts of the native areas (Houtman et al. 2004). To a great extent, *M. aquifolium* shared the group of northern *M. repens,* indicating the close relationship of these two taxa. Furthermore, our analyses indicate gene flow between the two taxa in the area of range-overlap. *M. aquifolium* and *M. repens* had been considered to be conspecific but later were accepted as two species (Piper 1906). Although both species apparently possess different morphological traits in habit, leaf colour and branching, they are sometimes difficult to distinguish and no single character can unambiguously identify either species (Piper 1922). *M. aquifolium* and *M. repens* hybridise not only in culture but also in nature (Houtman et al. 2004). This is confirmed by the admixture of the *M. repens* group in *M. aquifolium* in the STRUCTURE analysis. In particular the *M. aquifolium* population "na1" was clustered mostly with northern *M. repens,* which is based on shared alleles of this population to both groups (data not shown). In fact, the two species have overlapping ranges (Figure 1) and all populations included in our study originated from the sympatric range where the gene pools were not well separated. This may be either due to current gene flow between the species, but may also indicate an intermediate state of ongoing speciation within the *M. aquifolium* / *M. repens* group. The close relationship of *M. aquifolium* and *M. repens* complicated the analysis of the relationships between invasives, cultivars and native species. In particular the role of hybridisation in invasive *Mahonia* could hardly be detected because native *M. aquifolium* did not represent a well characterised uniform group. Rather, it was found to have an intermediate position between the *M. repens / M. pinnata* group and the group of invasive *Mahonia /* cultivars which were identified in the best supported STRUCTURE analysis ($K = 2$).

3.5.3   Cultivars as likely sources of invasive populations

We analysed a large number of cultivars, many of which were not assigned to the native species they were labeled by breeders (Figure 4 and Table 2). This may be due to recent gene flow. Thus, even the "pure" cultivated plants may descend from former hybridisation events, either in the native area or in cultivation. In cultivars the history and identity of introduced and bred individuals is often not traceable. There are several examples in which specimens were named erroneously (Piper 1922) or in which hybrid cultivars were named after one maternal species (Houtman et al. 2004). Ahrendt (1961) noticed that plants designated as *M. aquifolium* vary in morphology and consist largely of hybrids. Cultivated *M. aquifolium* produce stolons (Günther 1979) which actually is typical for native *M. repens* (Ahrendt 1961; Piper 1922), and indicates a hybrid origin of cultivated *M. aquifolium* plants. However, we could only detect a small proportion of the *M. repens* gene pool in some cultivars and hardly any in invasive populations.

We showed that hybridisation of *M. pinnata* and *M. aquifolium* seems to play a larger role for the cultivars than the hybridisation of *M. repens* and *M. aquifolium*. In view of the selection times of different cultivars, hybridisation with *M. pinnata* started earlier than hybridisation with *M. repens*. Further on, some of the old cultivars served as basis to breeding of other cultivars. *M. x wagneri* 'Moseri', for instance, served as basis for selection of *M. x wagneri* 'Sunset' (Houtman et al. 2004). Thus, genetic traits of older *M. pinnata* cultivars could infiltrate the population of cultivars. The age of a cultivar may play a role for the distribution across tree nurseries and garden centres and is therefore important for secondary release. High presence of alien plant species in trade and in consequence high planting rates enhance the likelihood of establishment in nature by enhanced propagule pressure (Kowarik 2005). In British nurseries more established alien plants are offered than casual alien species (Dehnen-Schmutz et al.

2007). Although the *M. pinnata* gene pool in cultivars should be old, we could not prove genes from *M. pinnata* in invasive populations, indicating that not all cultivars are similarly invasive. We could not prove significant *M. repens* proportions in invasive individuals either. Nevertheless, invasive *Mahonia* are different to native *M. aquifolium*.

According to our results, we assume that the differentiation of invasive *Mahonia* and native *M. aquifolium* is a result of different stages in the invasion process where restriction of gene pools, genetic drift, artificial selection and hybridisation had interacted. Thus, although multiple introductions are generally common (Bossdorf et al. 2005) and play a role also in the introduction of *Mahonia*, it is likely that genetic variability was actually reduced due to the selective import. Breeding and hybridisation could have enhanced genetic variability again (Ellstrand and Schierenbeck 2000). Simultaneously, genetic makeup was likely changed directed by artificial selection, which may have caused a bias in invasive founder populations (Kitajima et al. 2006). Thus, we assume that plant breeding facilitated invasion success of *Mahonia* by enhancing genetic variability and by generating characteristics that enhance invasiveness of certain cultivars.

Hybridisation may result in polyploid genotypes and fixed heterosis, that may fosters plant invasion (Ellstrand and Schierenbeck 2000). However, we did not find any evidence for polyploidy in invasive *Mahonia*, as only mono- or bi-allelic microsatellite genotypes were detected at all loci over all populations. Nevertheless, in other species polyploidy after hybridisation may play a role for successful invasion, for instance in *Spartina anglica* (Gray 1986).

3.5.4 Plant breeding and evolution of invasive traits

Recently, some plant traits have been identified, which may contribute to the invasion success of certain species in particular environments, even though no characteristic could be found that answered the basic question for invasive characteristics satisfactorily (Pysek et al. 1995; Lodge 1993). One characteristic that is known to enhance plant invasion is high seed production (Rejmanek 1996). This may play a role in invasive *Mahonia* populations, also. Many cultivars are praised for their large flowers, numerous fruits or the long residence time of berries at the sprout (Houtman et al. 2004). Soldaat and Auge (1998) suggested a horticultural effort for more flowers and fruits in *Mahonia*. Also other attributes, which plant breeders selected for in cultivated *Mahonia*, may be advantageous in natural habitats as well as in gardens. Invasive populations grow by vegetative below-ground stolons and stem layering (Auge and Brandl 1997), which is less known from *M. aquifolium* but from *M. repens* (Ahrendt 1961) and in some cultivars (Houtman et al. 2004). Thus, although we detected hardly any *M. repens* gene pool in invasive populations, clonal growth does obviously play an important role in these invasive populations. The cultivar *M. aquifolium* 'Maqu' (c14) is characterised especially by cold resistance, and a large amount of berries that are retained during autumn (Houtman et al. 2004), thereby facilitating seed dispersal by birds. *M. aquifolium* 'Maqu' is one old cultivar that shared a large proportion of genes with invasive populations and is possibly one of the successful invading cultivars.

These examples show that plant breeding may enhance invasion success by selecting for certain characteristics. Hence, there is more need to study characteristics of cultivated plant species and the relation of certain traits with invasiveness to identify general breeding efforts that go along with invasiveness. Breeders select especially for reproductive versatility, improvement in stress tolerance and pathogen resistance (Bundesverband Deutscher Pflanzenzüchter e.V. 2007), and broaden the phenotypic variation in particular by hybridisation.

These attributes will enable horticultural species to invade natural habitats. There are hardly any studies that investigated characteristics of horticultural plants in relation to invasiveness. Solely, Kitajima et al. (2006) showed that invasive *Ardisia crenata* individuals that descended from cultivars produce a greater number of seeds compared to native individuals. Large inflorescences, large flowers and high fruit production increase the number of seeds and may enhance invasiveness (Rejmanek 1996). Large numbers of seeds increases propagule pressure, especially in species that are bird-dispersed like *M. aquifolium*. Moreover, the birds which feed on *M. aquifolium* are common (e.g. blackbirds) (Torrey and Gray 1838), widespread and use both natural and urban habitat. Furthermore they are comparatively large birds which may further enhance dispersal distance. However, beside special traits that are selected by breeding and may enhance invasiveness, plant breeding may enhance invasion success simply by a mass effect: horticultural non-indigenous species become very abundant in a vast array of locations. Thus, the propagule pressure to adjacent native vegetation is greatly increased (Okada et al. 2007). Also, the many locations in which the species are grown differ in ecological conditions, which will increase the probability to find suitable conditions. In general, after first introduction, non indigenous species have been shown to undergo a lag phase before becoming invasive (Kowarik 1995). This time lag is hypothesised to be related to microevolution and local adaptation (Richards et al. 2006). Plant breeding and horticultural selection may shorten this lag phase by artificial selection of highly fecund genotypes and by distributing the cultivars and enhancing the probability of an invasion.

References

Ahrendt, L.W.A. (1961) *Berberis* and *Mahonia*. A taxonomic revision. *Journal of the Linnean Society of London, Botany,* **57**: 1-410.

Allendorf, F.W. and Lundquist, L.L. (2003) Introduction: Population Biology, Evolution, and Control of Invasive Species. *Conservation Biology,* **17**: 24-30.

Auge, H. and Brandl, R. (1997) Seedling recruitment in the invasive clonal shrub, *Mahonia aquifolium* Pursh (Nutt.). *Oecologia,* **110**: 205-211.

Barrett, S.C.H., and Richardson, B.J. (1986) Genetic attributes of invading species. In: *Ecology of biological invasions* (R.H. Groves and J.J. Burdon, eds), pp. 21-33 Cambridge: Cambridge University Press.

Bossdorf, O., Auge, H., Lafuma, L., Rogers, W.E., Siemann, E. and Prati, D. (2005) Phenotypic and genetic differentiation between native and introduced plant populations. *Oecologia,* **144**: 1-11.

Bundesverband Deutscher Pflanzenzüchter e.V. (2007) Züchtung. http://www.bdp-online.de/index.php?menu=3.

Dehnen-Schmutz, K., Touza, J., Perrings, C. and Williamson, M. (2007) The horticultural trade and ornamental plant invasions in Britain. *Conservation Biology,* **21**: 224-231.

Dieringer, D. and Schlötterer, C. (2002) Microsatellite analyser (MSA): a platform independent analysis tool for large microsatellite data sets. *Molecular Ecology Notes,* **3**: 167-169.

Durka, W., Bossdorf, O., Prati, D. and Auge, H. (2005) Molecular evidence for multiple introduction of garlic mustard (*Alliaria petiolata*, Brassicaceae) to North America. *Molecular Ecology*, **14**: 1697-1706.

Ellstrand, N.C. and Schierenbeck, K.A. (2000) Hybridization as a stimulus for the evolution of invasiveness in plants. *Proceedings of the National Academy of Sciences of the United States of America*, **97**: 7043-7050.

Evanno, G., Regnaut, S. and Goudet, J. (2005) Detecting the number of clusters of individuals using the software structure: a simulation study. *Molecular Ecology*, **14**: 2611-2620.

Goudet, J. (1995) FSTAT (Version 1.2): A computer program to calculate *F*- statistics. *Journal of Heredity*, **86**: 485-486.

PCAGEN 1999. Version 1.2. http://www2.unil.ch/popgen/softwares/pcagen.htm.

Gray, A.J. (1986) Do invading species have definable genetic characteristics? *Philosophical Transactions of The Royal Society of London*, **314**: 655-674.

Günther, H. (1979) Schöne Blütengehölze, Berlin: VEB Deutscher Landwirtschaftsverlag DDR.

Hollingsworth, M.L., Hollingsworth, P.M., Jenkins, G.I., Bailey, J.P. and Ferris, C. (1998) The use of molecular markers to study patterns of genotypic diversity in some invasive alien *Fallopia* spp. (Polygonaceae). *Molecular Ecology*, **7**: 1681-1691.

Houtman, R.T., Kraan, K.J. and Kromhout, H. (2004) *Mahonia aquifolium, M. repens, M. x wagneri* en hybriden. *Dendroflora*, **41**: 42-69.

Kim, Y.-D., Kim, S.-H. and Landrum, L.R. (2004) Taxonomic and phytogeographic implications from ITS phylogeny in *Berberis* (Berberidaceae). *Journal of Plant Research*, **117**: 175-182.

Kitajima, K., Fox, A.M., Sato, T. and Nagamatsu, D. (2006) Cultivar selection prior to introduction may increase invasiveness: evidence from *Ardisia crenata. Biological Invasions*, **8**: 1471-1482.

Klotz, S., Kühn, Ingolf, and Durka, Walter. (2002) BIOLFLOR - Eine Datenbank mit biologisch-ökologischen Merkmalen zur Flora von Deutschland. *Schriftenreihe für Vegetationskunde*, Vol. 38, Bonn - Bad Godesberg: Bundesamt für Naturschutz.

Kowarik, I. (1992) Einführung und Ausbreitung nichteinheimischer Gehölzarten in Berlin und Brandenburg. *Verhandlungen Botanischer Vereine Berlin Brandenburg*, **3**: 1-188.

Kowarik, I. (1995) Time lags in biological invasions with regard to the success and failure of alien species. In: *Plant Invasions - General Aspects and Special Problems* (P.

Pysek, K. Prach, M. Rejmanek and M. Wade, eds), pp. 15-38 Amsterdam: Academic Publishing.

Kowarik, I. (2005) Urban ornamentals escaped from cultivation. In: *Crop Ferality and Volunteerism* (J. Gressel, eds), pp. 97-121 Boca Raton: CRC Press.

Laferriere, J. (1997) Transfer of specific and infraspecific taxa from *Mahonia* to *Berberis* (Berberidaceae). *Botanicheskii Zhurnal*, **82**: 95-99.

Lodge, D.M. (1993) Biological Invasions: Lessons for Ecology. *Trends in Ecology & Evolution*, **8**: 133-137.

Mack, R.N. (2000) Cultivation fosters plant naturalization by reducing environmental stochasticity. *Biological Invasions*, **2**: 111-122.

Maron, J.L., Vila, M., Bommarco, R., Elmendorf, S. and Beardsley, P. (2004) Rapid evolution of an invasive plant. *Ecological Monographs*, **74**: 261-280.

Milne, R.I. and Abbott, R.J. (2000) Origin and evolution of invasive naturalized material of *Rhododendron ponticum* L. in the British Isles. *Molecular Ecology*, **9**: 541-556.

Monzingo, H.N. (1987) Shrubs of the Great Basin, Reno: University of Nevada Press.

Munz, P. (1959) A California Flora, Berkeley: University of California Press.

Neuffer, B., Auge, H., Mesch, H., Amarell, U. and Brandl, R. (1999) Spread of violets in polluted pine forests: morphological and molecular evidence for the ecological importance of interspecific hybridization. *Molecular Ecology*, **8**: 365-377.

Okada, M., Ahmad, R. and Jasieniuk, M. (2007) Microsatellite variation points to local landscape plantings as sources of invasive pampas grass (*Cortaderia selloana*) in California. *Molecular Ecology*, **16**: 4956-4971.

Perrins, J., Fitter, A. and Williamson, M. (1993) Population biology and rates of invasion of three introduced *Impatiens* species in the British Isles. *Journal of Biogeography*, **20**: 33-44.

Piper, C.V. (1906) Flora of the state of Washington. *Contributions from the United States National Herbarium*, **11**: 282-283.

Piper, C.V. (1922) The identification of *Berberis aquifolium* and *Berberis repens*. *Contributions from the United States National Herbarium*, **20**: 437-451.

Preston, C.D., Telfer, M.G., Arnold, H.R., Carey, P.D., Cooper, J.M., Dines, T.D., Hill, M.O., Pearman, D.A., Roy, D.B. and Smart, S.M. (2002) The changing flora of the UK, London: DEFRA: Oxford University Press.

Pritchard, J.K., Stephens, M. and Donnelly, P. (2000) Inference of population structure using multilocus genotype data. *Genetics,* **155**: 945-959.

Pysek, P., Prach, K., and Smilauer, P. (1995) Relating invasion success to plant traits: An analysis of the Czech alien flora. In: *Plant Invasions - General Aspects and Special Problems* (P. Pysek, K. Prach, M. Rejmanek and M. Wade, eds), pp. 39-60 Amsterdam: SPB Academic Publishing.

Rejmanek, M. (2000) Invasive plants: approaches and predictions. *Austral Ecology,* **25**: 497-506.

Rejmanek, M. (1996) A theory of seed plant invasiveness: The first sketch. *Biological Conservation,* **78**: 171-181.

Richards, C.L., Bossdorf, O., Muth, N.Z., Gurevitch, J. and Pigliucci, M. (2006) Jack of all trades, master of some? On the role of phenotypic plasticity in plant invasions. *Ecology Letters,* **9**: 981-993.

Rosenberg, N.A. (2004) DISTRUCT: a program for the graphical display of population structure. *Molecular Ecology Notes,* **4**: 137-138.

Roß, C. and Durka, W. (2006) Isolation and characterization of microsatellite markers in the invasive shrub *Mahonia aquifolium* (Berberidaceae) and their applicability in related species. *Molecular Ecology Notes,* **6**: 948-950.

Arlequin Version (2.000): a software for population genetics data analysis 2000. Switzerland: Genetics and Biometry Laboratory University of Geneva.

Soldaat, L.L., and Auge, H. (1998) Interactions between an invasive plant, *Mahonia aquifolium*, and a native phytophagous insect, *Rhagoletis meigenii*. In: *Plant Invasions: Ecological mechanisms and human responses* (U. Starfinger, K. Edwards, I. Kowarik and M. Williamson, eds), pp. 347-360 Leiden: Backhuys Publishers.

Torrey, J., and Gray, A. (1838) Berberidaceae. In: *Flora of North America* Oxford: Oxford University Press.

van de Laar, H.J. (1975) *Mahonia* en *Mahoberberis*. *Dendroflora,* **11/12**: 19-33.

Weir, B.S. and Cockerham, C.C. (1984) Estimating *F*-Statistics for the Analysis of Population-Structure. *Evolution,* **38**: 1358-1370.

Whittemore, A.T. (1997) *Berberis*. In: *Flora of North America* pp. 276-286 New York, Oxford: Oxford University Press.

Wright, S. (1951) The genetical structure of populations. *Annals of Eugenics,* **15**: 323-354.

Chapter 4

Invasive *Mahonia* plants outgrow their native relatives

In cooperation with Harald Auge

4.1 Abstract

Invasive populations often grow more vigorously than conspecific populations in the native range. This has frequently been attributed to evolutionary changes resulting either from founder effects, or from natural selection owing to enemy release. Another mechanism contributing to evolutionary change has largely been neglected in the past: Many invasive plant species do actually descend from cultivated plants and were therefore subject to breeding, including hybridisation and artificial selection. In a common garden experiment, we compared invasive Central European populations of the ornamental shrub, *Mahonia*, with native populations of its putative parental species, *Mahonia aquifolium* and *M. repens*, from North America. We hypothesised that plants of invasive populations show increased growth and retained high levels of heritable variation in phenotypic traits. Indeed, invasive *Mahonia* plants grew larger in terms of stem length, number of leaves and above-ground biomass than either of the two native species, which did not differ significantly from each other. Since there are no hints on release of invasive *Mahonia* populations from natural enemies, it is likely that hybridisation and subsequent selection by breeders have lead to an evolutionary increase of plant vigour in the introduced range. Further on, heritable variation was not consistently reduced in invasive populations compared with populations of the two native species. We suggest that interspecific hybridisation among the *Mahonia* species has counteracted the harmful effects of genetic bottlenecks often associated with species introductions. Based on this case study, we conclude that much more attention has to be paid on the role of plant breeding when assessing the mechanisms behind successful plant invasions in future.

4.2   Introduction

Invasive plant populations are often reported to perform better than conspecific populations in the native area (Blossey and Noetzold 1995; Crawley 1987). This difference between invasive and native populations may result from a more benign environment (Crawley 1987) but may also be genetically based. Genetic differentiation among native and invasive populations in neutral markers as well as in quantitative traits has been shown for several plant species (reviewed in Bossdorf et al. 2005; e.g. Durka et al. 2005; Blair and Wolfe 2004; DeWalt and Hamrick 2004) and can be caused by random genetic drift, hybridisation and gene flow in the new area, or by evolutionary adjustments to the novel environment (Allendorf and Lundquist 2003). Genetic adaptations may influence both the number of invaded habitats and the dominance within habitats (Parker et al. 2003), and there is increasing evidence that the ability for adaptive evolution is a key feature of successful invaders (Bossdorf et al. 2005; Sakai et al. 2001; Caroll and Dingle 1996). However, one factor that has been uncared in invasion biology for a long time is the role of plant breeding, even though a high proportion of successful alien plants descent from cultivated plants (Kuehn and Klotz 2003; Preston et al. 2002). The role of cultivation in plant invasions has been mainly discussed with respect to increased propagule pressure (Mack 2000), but genetic changes due to plant breeding may contribute to invasion success as well: First, artificial selection for traits that are of ornamental value but also ecologically relevant may enhance the chance of naturalisation and spread (Kitajima et al. 2006). Second, breeding often involves hybridisation with related species, which is suggested to facilitate the evolution of invasiveness by enhancing genetic variation, creating novel genotypes, dumping of genetic load, and by fixed heterosis effects (Ellstrand and Schierenbeck 2000). Hybridisation may thus be a mechanism compensating for genetic bottlenecks which are commonly associated with species introductions and are known to impede evolutionary adjustments (Allendorf and Lundquist 2003). On the other hand, in many cases hybrids spread by extensive vegetative proliferation and thus, even populations with low genetic variation may become serious invaders (Vila et al. 2000). Hence, more attention has to be paid to the potential contribution of man-made genetic changes to the invasion success of ornamental or other cultivated plants.

Using neutral molecular markers, several studies have shown that invasive taxa originated by hybridisation of exotic species with native species (Moody and Les 2002; Ayres et al. 1999) or other exotic species (Gaskin and Schaal 2002; Hollingsworth et al. 1998). Only a few studies, however, compared invasive hybrids with their native parents in ecological relevant, phenotypic traits (but see Facon et al. 2005; Vila and D'Antonio 1998). In our paper, we compare invasive populations of an ornamental shrub with its putative parental species from the native area with respect to plant growth and within-population genetic variation. *Mahonia aquifolium* (PURSH) NUTT. is native in North America and was introduced to Europe in the 19th century (Hayne 1822, cited in Kowarik 1992) as an ornamental plant. Plants cultivated today under the name *M. aquifolium* are largely casual hybrids, probably mainly with the closely related *M. repens* (LINDL.) G. DON (Ahrendt 1961). In addition, many cultivars deliberately bred by hybridisation of the two species are cultivated (Houtman et al. 2004). *M. repens* is used as an ornamental plant, too, and *M. repens* as well as the hybrids are known to escape from cultivation (Clement and Foster 1994; Stace 1991). In Central Europe, descendants of these cultivated forms are successful invaders of a wide range of habitats, ranging from shady forest to open, dry scrub (Auge and Brandl 1997; Kowarik 1992; Lohmeyer and Sukopp 1992). Although the actual genetic status of the invasive *Mahonia* populations has not been proved yet, it is most likely that invasive populations represent a hybrid swarm between the two species. We carried out a common garden experiment with invasive *Mahonia* populations from Central Europe, and native *M. aquifolium* and *M. repens* populations from North America, to test the hypotheses that

(1)   invasive populations show increased plant size relative to the putative parental species, and

(2)   invasive populations show the same magnitude of heritable variation in quantitative traits like native populations.

4.3   Material and Methods

4.3.1   Study species

Mahonia aquifolium and related species are native to western North America. They are shrubs with evergreen, pinnate leaves, insect-pollinated flowers, and vertebrate-dispersed berries. *M. aquifolium* is characterised by simple and erect stems reaching 1.80 m in height (Ahrendt 1961; Piper 1922). It occurs from British Columbia to California and from of the Pacific coast to Montana and Idaho, and grows in open woods and shrublands from sea level to 2100 m altitude (Whittemore 1997). The closely related *M. repens* reaches 90 cm in height and is more stoloniferous than *M. aquifolium* (Ahrendt 1961). Although there is some overlap in distribution, *M. repens* occurs more to the north, south and east than *M. aquifolium*. It grows in open forests, shrublands and grasslands up to an altitude of 3000 m (Whittemore 1997). *M. aquifolium* was introduced to Central Europe for ornamental purpose in 1822 (Hayne 1822, cited in Kowarik 1992) and repeatedly later on. Breeding resulted in a large number of cultivars and involved hybridisation with related species (see van de Laar 1975). Cultivars and stocks that are probably mainly hybrids between *M. repens* and *M. aquifolium* (see Ahrendt 1961), have been frequently planted at road sides, in gardens and in parks, and are known to escape from cultivation (Clement and Foster 1994; Stace 1991). The first spontaneous occurrence outside gardens was observed in 1860 after a time lag of 38 years (Kowarik 1992). Today, the descendants of the cultivated forms are successful neophytes in Central Europe invading anthropogenic as well as natural vegetation (Kowarik 1992; Lohmeyer and Sukopp 1992). While regional spread depends on seed dispersal by vertebrates, local population increase takes place by repeated seedling recruitment and extensive clonal growth (Auge and Brandl 1997; Auge et al. 1997). Because of the hybridisation involved, we will use the term invasive *Mahonia* populations in this article to distinguish them from the other two taxa, i.e. their putative parental species, *M. aquifolium* and *M. repens*. Invasive *Mahonia* populations show high phenotypic variation (Auge et al. 1997) and large variation in neutral genetic markers (Ross et al. submitted).

4.3.2   Sampling and rearing of plant material

Berries of 13 invasive European *Mahonia* populations, four native *M. aquifolium* populations, and five native *M. repens* populations were collected in summer 2003 (Table 1). The offspring of each maternal plant represents at least half-sibs and will therefore be referred to as a seed family. In December 2003, we sowed seeds of five to eight families of each population in plastic trays containing a mixture of 50 % rearing compost (Composana; COMPO GmbH, Münster, Germany) and 50 % sand, and stored them at 5 °C in a refrigerator. After 16 weeks of stratification, we transferred the seed trays to a 14 h/10 h day/night cycle at 15 °C/10 °C. Three weeks later we raised the day temperature to 20 °C to facilitate germination. After further two weeks, we planted the seedlings separately in 3 litre plastic pots of 12 cm diameter with 50 % of a standard potting soil (Fruhstorfer Typ P, florimaris Humus- und Erdenwerk GmbH & Co. KG, Wangerland, Germany) and 50 % sand, and placed them in a greenhouse with a day/night cycle of 14 h/10 h and 25 °C/15 °C until the start of the experiments.

Table 1: Sampling locations of native *Mahonia aquifolium* and *M. repens* populations, and of invasive *Mahonia* populations used in the common garden experiment.

Status	Taxon	Location	Geographic coordinates
native	*M. aquifolium*	British Columbia: Tie Lake	49.3 N 115.2 W
native	*M. aquifolium*	Idaho: Harvard	46.4 N 117.0 W
native	*M. aquifolium*	Oregon: site A	45.2 N 118.1 W
native	*M. aquifolium*	Oregon: site B	44.4 N 123.3 W
native	*M. repens*	Montana: Blackfoot River	45.6 N 113.4 W
native	*M. repens*	Montana: Boulder River	45.2 N 110.1 W
native	*M. repens*	Montana: Rattlesnake Mountain	46.5 N 113.6 W
native	*M. repens*	Idaho: Deary	46.5 N 116.3 W
native	*M. repens*	Idaho: Deer Road	46.5 N 116.0 W
invasive	invasive *Mahonia*	Germany: Barby	51.6 N 11.6 E
invasive	invasive *Mahonia*	Germany: Buckow	52.3 N 14.1 E
invasive	invasive *Mahonia*	Germany: Drebkau	51.4 N 14.1 E
invasive	invasive *Mahonia*	Germany: Duebener Heide	51.4 N 12.3 E
invasive	invasive *Mahonia*	Germany: Herzfelde	52.3 N 13.5 E
invasive	invasive *Mahonia*	Germany: Kirchbrak	51.6 N 9.4 E
invasive	invasive *Mahonia*	Germany: Halle-Lieskau	51.3 N 11.6 E
invasive	invasive *Mahonia*	Germany: Liepe	52.5 N 13.6 E
invasive	invasive *Mahonia*	Germany: Linz am Rhein	50.3 N 7.2 E
invasive	invasive *Mahonia*	Germany: Mannheim	49.3 N 8.3 E
invasive	invasive *Mahonia*	Germany: Neuhaus an der Pegnitz	49.4 N 11.3 E
invasive	invasive *Mahonia*	Czech Republik: Prague	50.7 N 14.3 E
invasive	invasive *Mahonia*	Germany: Rothenburg	51.4 N 11.5 E

To account for differences in initial seedling size which may reflect maternal environmental effects, we determined the length of the longest leaf of all seedlings at the start of the experiment. According to a preliminary test using 50 *Mahonia* seedlings, the length of longest leaf proved to be the best predictor of seedling biomass ($r^2 = 0.28$) compared to other non-destructive measures.

4.3.3   Experimental design and measurements

In July 2004, we set up a common garden experiment at the UFZ Experimental Station in Bad Lauchstädt, Germany (51.37 ° N, 11.83 ° E), using a split-plot design. The main plot level consisted of six plots and was used to investigate the effects of shading: Three randomly chosen plots received full sunlight, whereas three other plots were covered with neutral PE shade cloth and received only about 55 % of light. We established two different light treatments, because native *M. repens* occur in more open habitats in contrast to native *M. aquifolium,* and we wanted to avoid inappropriate light conditions for one of these species. In addition, this main plot treatment allowed us to compare the response of the three taxa to different light conditions. The subplot level consisted of individual plants and was used to compare the taxa, populations and seed families. One seedling of each seed family was randomly positioned within each plot, except some seed families with poor germination that could not be distributed across all plots. The pots were embedded with bark mulch to protect them from extreme temperatures and were watered when necessary.

At the end of October 2005, we terminated the experiment and harvested all plants. We measured three functional traits of each plant: chlorophyll content, specific leaf area (SLA) and leaf area ratio (LAR). Chlorophyll content and SLA were quantified using three randomly chosen leaves per plant. To quantify chlorophyll content, we used a chlorophyll meter (SPAD-502, Minolta Co. Ltd., Osaka, Japan) which measures the amount of light transmitted by the leaf in two specific wavelength regions. Using 30 additional *Mahonia* individuals we found that chlorophyll content measured with the SPAD-502, and chlorophyll concentration determined according to Lichtenthaler & Wellburn (1983) were highly correlated ($r^2 = 0.79$). It should be noted, however, that the SPAD measures chlorophyll content on an arbitrary scale and on the basis of leaf area rather than leaf mass. To quantify SLA, we determined leaf area using a LI-3100 Area Meter (LI-COR Biocience, Lincoln, NE, USA) before drying the leaves at 80 °C. We calculated LAR in terms of leaf area per unit above-ground biomass and SLA as expression of leaf area per unit leaf mass.

To characterise plant size, we counted the number of stems and leaves, measured the length of longest stem, determined the biomass of leaves and stems after drying at 80 °C, and calculated total aboveground biomass. We did not consider reproductive traits because only few plants had flowered, and belowground structures because the inappropriate texture of potting soil prevented a reliable measurement of root biomass.

4.3.4 Statistical analyses

Data were analysed according to a split-plot design with treatment on the plot level, and taxon, population and seed family on the subplot level. For the functional traits (chlorophyll content, SLA and LAR) we calculated an ANOVA, and for the size-related traits an ANCOVA with length of longest leaf at the start of the experiment as covariate to account for initial seedling size (PROC GLM, SAS version 9.1, SAS Institute, Cary, NC, USA). Because of poor germination and subsequent mortality not all seed families and populations were still present in each treatment at harvest time, and the effects of seed family and population were thus confounded with the shading effect. Therefore, we calculated type I sum of squares and fitted shading and plot nested within shading first to produce robust results for the effects of taxon, population and seed family. We considered population, seed family and their interactions with shading as random effects. Therefore, taxon was tested against population within taxon, population against seed family within population, shading x taxon against shading x population within taxon, and shading x population within taxon against shading x seed family within population. All other effects were tested against the residuals. We compared the last square means of the three taxa by GT2-method (Hochberg 1974), that is recommended for unequal sample size (Sokal and Rohlf 1995). Since we intended to avoid inflated type II errors as result of Bonferroni tests (Cabin and Mitchell 2000), we calculated the probability to find one significant result by chance which is $1 - 0.95^3 = 0.14$ in the three functional traits, and $1 - 0.95^6 = 0.26$ in the six size-related traits (Moran 2003). SLA, LAR, stem length, number of leaves and all biomass data were log transformed, and number of stems was square-root transformed to approach normality and homoscedasticity. For a comparison of the whole set of plant traits between the three taxa we calculated a principal component analysis on chlorophyll content, SLA, LAR, stem length, number of leaves, number of stems, leaf biomass and stem biomass using least square means for populations (PROC PRINCOMP).

To compare the heritable component of variation between the three taxa, we estimated variance components between and within seed families for all traits (untransformed data) using the restricted maximum likelihood method (PROC MIXED). Variance components were calculated for each population separately. Since we had no information whether seed families represent half-sibs or full-sibs, we used the intraclass correlation coefficient as a measure for heritable variation: $t = \sigma_{bf}^2 / (\sigma_{bf}^2 + \sigma_{wf}^2)$ with σ_{bf}^2 being the variance between seed families and σ_{wf}^2 the variance within seed families. This coefficient describes the resemblance among relatives and is thus a measure of genetic variation (Falconer and Mackay 1996; Lawrence 1984). Intraclass correlation coefficients were compared between the three taxa using ANOVA and GT2-tests.

4.4   Results

Whereas all traits varied strongly among plots, shading did not significantly affect any plant trait despite chlorophyll content (Tables 2 and 3). Chlorophyll content (as measured on leaf area basis) was reduced by 9 % in shade compared to full sunlight, which might result from a slight but non-significant increase in SLA. Responses to shading did neither vary among the three taxa nor among populations as indicated by the non-significant shading x taxon and shading x population interactions, respectively.

Out of the three functional traits measured, only chlorophyll content differed significantly between the three taxa (Table 2). Although native *M. aquifolium* had slightly lower chlorophyll content as well as slightly higher SLA and LAR compared to native *M. repens* and invasive *Mahonia* (Table 4), no pairwise comparison between the three taxa was significant. Furthermore, we did not detect significant differences of functional traits among populations within taxa and among seed families within populations.

Initial seedling size had an effect of all traits related to plant size at the end of the experiment (Table 3). The three taxa differed significantly in all these traits except in the number of stems (Table 4). The differences between native *M. aquifolium* and *M. repens* were not significant. Invasive *Mahonia* populations attained significantly larger values than both native species in all traits related to plant size except in the number of stems. Note that the significant differences were more frequent than expected by chance in a table with six comparisons. Populations within taxa varied significantly in number of leaves and number of stems, and seed families within populations differed in leaf, stem and total aboveground biomass (Table 3). In case of the among-population variation, however, these are just as many significant results as expected by chance.

Table 2: Results of analyses of variance for functional traits. Degrees of freedom are given for the effect and the respective error mean squares (in brackets: error degrees of freedom for chlorophyll content). Significance levels of F ratios are given as follows: $* p < 0.05$, $** p < 0.01$, $*** p < 0.001$.

	Degrees of freedom	*F* ratios		
		Chlorophyll content	Specific leaf area	Leaf area ratio
Shading	1, 4	9.48 *	2.41	1.72
Plot (shading)	4, 401 (404)	3.32 *	13.11***	6.52 ***
Taxon	2, 19	4.69 *	2.68	1.11
Population (taxon)	19, 100	1.39	0.99	1.56
Seed family (population x taxon)	100, 401 (404)	1.09	1.23	1.30
Shading x taxon	2, 18	0.97	0.48	0.59
Shading x population (taxon)	18, 76	0.77	1.59	0.79
Shading x seed family (pop x taxon)	76, 401 (404)	1.22	1.01	1.30

Table 3: Results of analyses of covariance for traits related to plant size. Degrees of freedom are given for the effect and the respective error mean squares (in brackets; error degrees of freedom for plant height). Significance levels of F ratios are given as follows: $* p < 0.05$, $** p < 0.01$, $*** p < 0.001$.

	Degrees of freedom	F ratios					
		Aboveground biomass	Biomass of stems	Biomass of leaves	Stem length	Number of stems	Number of leaves
Initial seedling size	1, 411 (406)	54.57 ***	57.70 ***	47.20 ***	51.46 ***	5.88 *	22.54 ***
Shading	1, 4	0.84	0.89	0.84	0.57	1.50	2.64
Plot (shading)	4, 411 (406)	26.25 ***	20.07 ***	24.61 ***	10.47 ***	8.70 ***	37.29 ***
Taxon	2, 19	15.21 ***	16.09 ***	12.64 ***	21.99 ***	2.82	15.05 ***
Population (taxon)	19, 100	1.02	1.02	1.05	0.78	1.70 *	1.75 *
Seed family (population x taxon)	100, 411 (406)	1.36 *	1.36 *	1.33 *	1.17	1.17	1.13
Shading x taxon	2, 18	1.20	1.35	1.01	0.64	1.25	2.07
Shading x population (taxon)	18, 76	1.01	1.02	0.90	0.88	1.06	0.59
Shading x seed family (pop x taxon)	76, 411 (406)	1.27	1.08	1.39 *	1.00	1.07	1.73 ***

In the multivariate analysis, the first three principal components together explained 91 % of the variance in plant traits (Figure 1). The first axis was positively associated with traits related to plant size and separated all three taxa well (*M. aquifolium*: - 0.89 ± 0.90, *M. repens*: -2.93 ± 0.32, invasive *Mahonia* populations: 1.40 ± 0.22). Hence, the position of the invasive populations along this axis reflects their increased growth compared to the putative parental species. The second axis was mainly associated with functional traits, i.e. positively with SLA and LAR, and negatively with chlorophyll content. It separated the two native species (*M. aquifolium*: 1.48 ± 0.54, *M. repens*: -0.83 ± 0.39), whereas the invasive *Mahonia* populations overlapped with both species and showed, on average, intermediate values (-0.14 ± 0.38). The third axis was positively associated with chlorophyll content and LAR, but did not differentiate the three taxa.

Table 4: Functional and size-related traits of native *M. aquifolium* and *M. repens* from North America, and invasive *Mahonia* populations from Central Europe, grown in a common garden experiment. Least square means and standard errors among populations within each taxon are given. Different superscript letters indicate significant differences at $p < 0.05$ according to GT2-tests (calculated on transformed data in the case of stem length, number of stems, number of leaves and all biomass data). In all size-related traits, except the number of stems, invasive *Mahonia* populations perform significantly better than the two native species.

	Native species		Invasive *Mahonia* populations
	M. aquifolium	*M. repens*	
Functional traits			
Chlorophyll content	36.66 ± 1.29 [a]	39.48 ± 2.16 [a]	40.38 ± 0.62 [a]
Specific leaf area [cm²/g]	91.21 ± 2.93 [a]	99.91 ± 5.13 [a]	91.14 ± 3.28 [a]
Leaf area ratio [cm²/g]	69.61 ± 2.71 [a]	66.23 ± 4.65 [a]	65.38 ± 1.31 [a]
Size-related traits			
Aboveground biomass [g]	1.95 ± 0.51 [a]	1.56 ± 0.89 [a]	3.65 ± 0.25 [b]
Biomass of stems [g]	0.48 ± 0.16 [a]	0.31 ± 0.28 [a]	0.94 ± 0.08 [b]
Biomass of leaves [g]	1.47 ± 0.36 [a]	1.25 ± 0.64 [a]	2.71 ± 0.18 [b]
Stem length [cm]	7.42 ± 0.62 [a]	5.72 ± 1.10 [a]	9.15 ± 0.31 [b]
Number of stems	3.90 ± 0.61 [a]	2.85 ± 1.08 [a]	5.03 ± 0.30 [a]
Number of leaves	12.08 ± 1.62 [a]	6.89 ± 2.85 [a]	17.61 ± 0.79 [b]

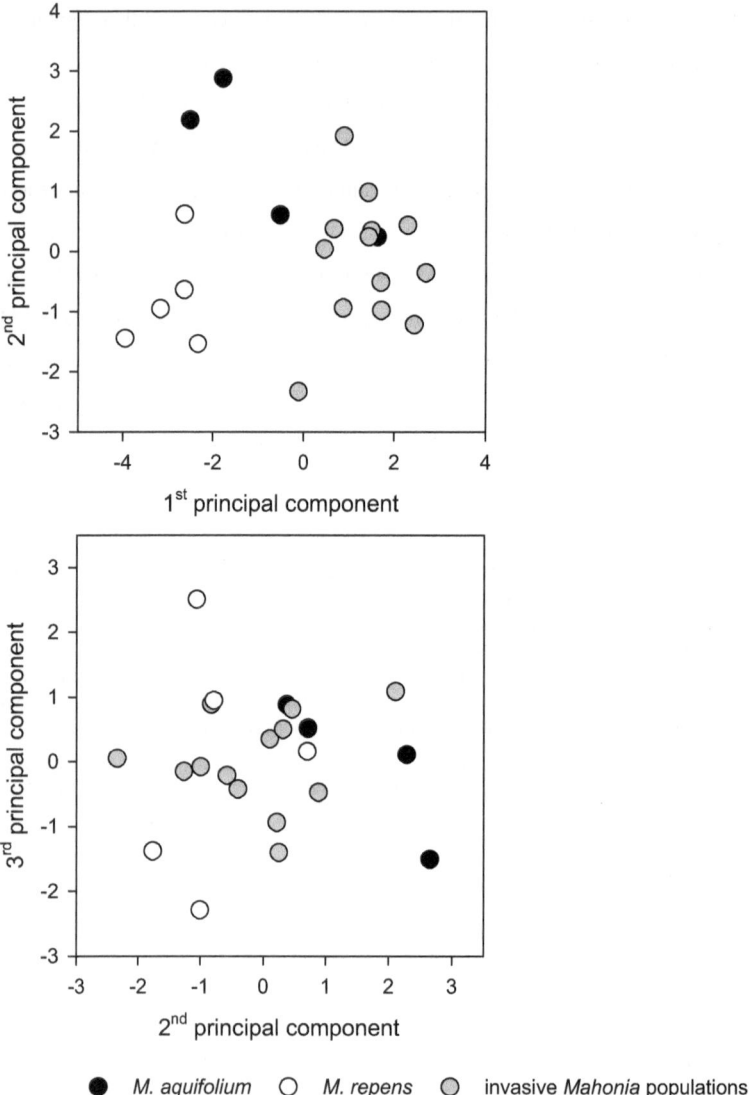

● *M. aquifolium*   ○ *M. repens*   ◉ invasive *Mahonia* populations

Figure 1: Results of principal component analysis of three functional and five size-related traits of four native *M. aquifolium* populations and five native *M. repens* populations from North America, and 13 invasive *Mahonia* populations from Central Europe, grown in a common garden experiment. The first three principal components explain 54 %, 24 % and 13 % of the variance in plant traits, respectively.

Intraclass correlation coefficients of functional and size-related traits of the three taxa ranged from 0 to 0.402 (Table 5) and were, averaged across taxa, highest in SLA (0.220) and lowest in stem length (0.039). Invasive *Mahonia* populations had a significantly lower heritable variation in all functional traits compared to *M. repens* but not to *M. aquifolium*, whereas the two native species did not differ significantly. In all size-related plant traits, we found no significant differences among intraclass correlation coefficients of the three *Mahonia* taxa.

Table 5: Heritable variation of quantitative traits within populations of native *Mahonia* species and within invasive *Mahonia* populations as measured by the intraclass correlation coefficient. Means and standard errors among populations within each taxon are given. Different superscript letters indicate significant differences at $p < 0.05$ according to GT2-tests.

	Intraclass correlation coefficient t		
	Native species		Invasive *Mahonia* populations
	M. aquifolium	*M. repens*	
Functional traits			
Chlorophyll content	0.014 ± 0.067 [ab]	0.249 ± 0.060 [a]	0.058 ± 0.037 [b]
Specific leaf area	0.216 ± 0.099 [ab]	0.402 ± 0.089 [a]	0.041 ± 0.055 [b]
Leaf area ratio	0.004 ± 0.044 [ab]	0.121 ± 0.039 [a]	0.000 ± 0.024 [b]
Size-related traits			
Aboveground biomass	0.005 ± 0.059 [a]	0.156 ± 0.053 [a]	0.049 ± 0.033 [a]
Biomass of stems	0.011 ± 0.047 [a]	0.067 ± 0.042 [a]	0.070 ± 0.026 [a]
Biomass of leaves	0.014 ± 0.063 [a]	0.168 ± 0.056 [a]	0.039 ± 0.035 [a]
Stem length	0.043 ± 0.038 [a]	0.000 ± 0.034 [a]	0.073 ± 0.021 [a]
Number of stems	0.056 ± 0.074 [a]	0.182 ± 0.066 [a]	0.051 ± 0.041 [a]
Number of leaves	0.089 ± 0.038 [a]	0.042 ± 0.034 [a]	0.027 ± 0.021 [a]

4.5 Discussion

With respect to our initial hypotheses, we summarise our results as follows:

(1)   Invasive *Mahonia* populations showed a strongly increased plant size relative to the putative parental species, *M. aquifolium* and *M. repens*.

(2)   Invasive *Mahonia* populations showed a similar magnitude of heritable variation in quantitative traits like native *Mahonia* populations.

Invasive *Mahonia* populations descend from cultivated stocks and are supposed to be largely of hybrid origin, most likely between *M. aquifolium* and *M. repens* (Ahrendt 1961). There are some other North American *Mahonia* species cultivated in Europe, in particular *M. pinnata* (Lag.) Fedde which is also known to be hybridised with *M. aquifolium* (see Houtman et al. 2004; van de Laar 1975). Although we still do not know the actual extent of gene flow among these species, our comparisons considers the two most likely parental species of invasive *Mahonia* populations. Genetic analyses using microsatellite markers (Roß and Durka 2006) reveal that invasive *Mahonia* populations are clearly separated from native *M. aquifolium*. They also show that *M. pinnata* genes are apparently not present in invasive populations, while the narrow relationship between *M. aquifolium* and *M. repens* has rendered it difficult to separate these two species and, hence, to detect hybridisation between them so far (Ross et al 2008).

Like the major part of studies comparing native versus introduced plant populations (see references in Bossdorf et al. 2005), our experiment was based on seeds produced by open pollination in the field rather than by controlled pollination under controlled conditions (cf. Falconer and Mackay 1996; Lawrence 1984). Phenotypic variation among taxa, populations and seed families has therefore a genetic and a maternal environmental component. Since the maternal environment often becomes manifest in seed provisioning and early seedling growth (Rossiter 1996; Roach and Wulff 1987), we used initial seedling size as covariate for size-related plant traits to account for such effects. For above-ground biomass, e.g., this reduces the variance attributed to differences among taxa, populations and seed families by 39 %, 31 % and 5 %, respectively. Although this approach controls for possible maternal effects it may underestimate genetic variation, if differences in seed or seedling size were mainly genetically controlled. Our comparisons are, thus, rather conservative and we feel confident that they do primarily reflect genetically based differences.

We also have to consider that our results refer specifically to the environmental conditions of our experimental garden. Maron et al. (2004) presented an experiment comparing invasive and native St. John's wort in four common gardens, and demonstrated that differences among populations were dependent on the particular environment. In order to take into account possible genotype x environment interactions at least with respect to light, we conducted our experiment under two different light conditions. Since we found no difference in the response of the three taxa, the superiority of invasive *Mahonia* populations is obviously not a result of favourable light conditions associated with enhanced phenotypic plasticity.

Our finding of an enhanced growth of invasive populations compared with native populations is in line with the patterns found in 56 % of the studies recently reviewed by (Bossdorf et al. 2005). These results were mostly interpreted in terms of the Evolution of Increased Competitive Ability (EICA) hypothesis: Invasive plants should be subject to a release from natural enemies in the new area (Keane and Crawley 2002), and should therefore evolve towards increased growth and reduced defence (Blossey and Noetzold 1995). However, we do not have any evidence for a release of invasive *Mahonia* populations from natural enemies. They are heavily attacked by *Rhagoletis meigenii* Loew, a specialised seed-predating fruit fly native to Europe (Soldaat and Auge 1998), by *Cumminsiella mirabilissima* (Pk.) Nannf., a specialised rust fungus introduced from North America, as well as by a native generalist leafhopper (*Aphrophora alnii* (Fallén) (Auge et al. 1997)). Furthermore, invasive *Mahonia* plants do not harbour a lower diversity and abundance of invertebrates than native woody species, and do not show an increased palatability (i.e., reduced defence) of leaves compared with populations from the native range (M. Brändle, C. Belle & R. Brandl, unpublished results). We therefore suggest the increased vigour of invasive *Mahonia* populations to be caused by plant breeding, involving both hybridisation and artificial selection, rather than by evolutionary responses to enemy release.

Interspecific hybridisation is rather common in plants (Arnold and Hodges 1995) and has recently been considered as a stimulus for the evolution of invasiveness (Ellstrand and Schierenbeck 2000; Vila et al. 2000). Superior growth of hybrids may be caused by heterosis, i.e. increased vigour based on high levels of heterozygosity in the F_1 (Lynch and Walsh 1998). Heterosis decays rapidly in subsequent generations but can become fixed, e.g., by allopolyploidy or clonal growth (Rieseberg et al. 2007; Ellstrand and Schierenbeck 2000). There is, however, no evidence for fixed heterosis in invasive *Mahonia* populations: we found no hint for allopolyploidy (C.A. Ross, unpublished data), and clonal growth is not the predominant mode of reproduction and spread (Auge and Brandl 1997). Since our invasive *Mahonia* plants were definitely not F_1 hybrids, it is therefore unlikely that heterosis as a transitory phenomenon had caused their increased size. However, hybridisation usually leads to increased genetic variation in the F_2 and later generations (Lynch and Walsh 1998) thereby creating novel genotypes and providing the raw material for evolutionary changes (Rieseberg et al. 2007; Ellstrand and Schierenbeck 2000). We suspect that early hybrid genotypes formed the raw material which was then used by plant breeders to select, either intentionally or unintentionally, for particularly vigorous growth. The physiological mechanism behind the increased vigour of invasive *Mahonia* plants, however, remains dubious because it was not related to any of the functional traits measured.

The evolutionary novelty of hybrids is frequently reflected by genotype x environment interactions, i.e. superior hybrid fitness in habitats novel to both parental species (Arnold and Hodges 1995). This may allow the colonisation of new habitats as it was repeatedly shown for hybrids among native species (e.g. Rieseberg et al. 2007; Neuffer et al. 1999). The spread of hybrids among introduced species, or between native and introduced species, can be considered as an ecologically and evolutionarily similar process: for instance, invasive *Reynoutria x bohemica* spreads faster than its two exotic parents, *R. japonica* and *R. sachaliniensis*, in Central Europe (Mandak et al. 2004), and occurs also in habitats were both parents are absent (Hollingsworth et al. 1998). There are an increasing number of studies showing that invasive plant taxa are essentially hybrids, e.g. *Tamarix* populations (Gaskin and Schaal 2002) and *Myriophyllum* populations (Moody and Les 2002) in North America, and *Rhododendron ponticum* in Europe (Milne and Abbott 2000). However, only few studies actually compared functional traits, plant growth, or fitness of invasive hybrids with that of their parental species. Exceptions are the studies on *Carpobrotus* hybrids in California which showed increased growth compared to either parental species (Vila and D'Antonio 1998) and *Rhododenron ponticum* which showed increased growth in comparison to one parent species from different origins (Erfmeier and Bruelheide 2005). *Rhododendron ponticum* originates from disjunctive areas in Europe, was intentionally hybridised with cold tolerant species from North America (Milne and Abbott 2000), and is one of the most serious weeds in Great Britain today (Cross 1982).

The two examples, *Rhododendron* and *Mahonia*, demonstrate that plant breeding, which involves hybridisation as a common technique, has to be recognised as a critical factor contributing to the evolution of invasive plant taxa.

The importance of hybridisation among different *Mahonia* species may reach even beyond the provision of novel, vigorously growing genotypes. Species introductions are usually assumed to involve genetic bottlenecks (Allendorf and Lundquist 2003) that will increase the likelihood of evolution by random drift (Barrett and Husband 1990) but – since response to selection is proportional to heritable variation (Lynch and Walsh 1998) – impede adaptive evolution. Moreover, bottlenecks can lead to inbreeding depression that may limit population growth and increase extinction risk (Allendorf and Lundquist 2003). In contrast to the expected bottlenecks, however, genetic variation of invasive plant populations is not generally lower than that of conspecific native populations (Bossdorf et al. 2005). Besides multiple introductions from different source populations like, e.g., in the invasive herb *Alliaria petiolata* (Durka et al. 2005), interspecific hybridisation may help introduced species to overcome harmful bottleneck effects (Ellstrand and Schierenbeck 2000). Indeed, we found heritable variation in ecologically relevant traits not being consistently reduced in invasive *Mahonia* populations compared with native populations. We think that hybridisation among the two native taxa, in addition to multiple introductions of both of them, has compensated for any harmful effect of genetic bottlenecks.

In conclusion, we were able to show that plants of invasive *Mahonia* populations are more vigorously growing than plants from native populations of both putative parental species. Most likely, increased growth results from hybridisation and artificial selection during plant breeding, because there is neither a hint on founder effects nor any indication for evolution towards increased growth at the expense of defence. Since a large proportion of invasive plant species descent from cultivated or ornamental plants (Kuehn and Klotz 2003; Preston et al. 2002; Reichard and Hamilton 1997), the role of plant breeding should receive much more attention in further studies.

References

Ahrendt, L.W.A. (1961) *Berberis* and *Mahonia*. A taxonomic revision. *Journal of the Linnean Society of London, Botany,* **57**: 1-410.

Allendorf, F.W. and Lundquist, L.L. (2003) Introduction: Population Biology, Evolution, and Control of Invasive Species. *Conservation Biology,* **17**: 24-30.

Arnold, M.L. and Hodges, S.A. (1995) Are natural hybrids fit or unfit relative to their parents? *Trends in Ecology & Evolution,* **10**: 67-71.

Auge, H. and Brandl, R. (1997) Seedling recruitment in the invasive clonal shrub, *Mahonia aquifolium* Pursh (Nutt.). *Oecologia,* **110**: 205-211.

Auge, H., Brandl, R. and Fussy, M. (1997) Phenotypic variation, herbivory and fungal infection in the clonal shrub *Mahonia aquifolium* (Berberidaceae). *Mitteldeutsche Gesellschaft für Allgemeine und Angewandte Entomology,* **11**: 747-750.

Ayres, D.R., Garcia-Rossi, D., Davis, H.G. and Strong, D.R. (1999) Extent and degree of hybridization between exotic (*Spartina alterniflora*) and native (*S. foliosa*) cordgrass (Poaceae) in California, USA determined by random amplified polymorphic DNA (RAPDs). *Molecular Ecology,* **8**: 1179-1186.

Barrett, S.C.H., and Husband, B.C. (1990) Genetics of plant migration and colonization. In: *Plant population genetics, breeding, and genetic resources* (A.H.D. Brown, M.T. Clegg, A.L. Kahler and B.S. Weir, eds), pp. 254-277 Sunderland: Sinauer.

Blair, A.C. and Wolfe, L.M. (2004) The evolution of an invasive plant: An experimental study with *Silene latifolia*. *Ecology,* **85**: 3035-3042.

Blossey, B. and Noetzold, R. (1995) Evolution of increased competitive ability in invasive nonindigenous plants: A hypothesis. *Journal of Ecology,* **83**: 887-889.

Bossdorf, O., Auge, H., Lafuma, L., Rogers, W.E., Siemann, E. and Prati, D. (2005) Phenotypic and genetic differentiation between native and introduced plant populations. *Oecologia,* **144**: 1-11.

Cabin, R.J. and Mitchell, R.J. (2000) To Bonferroni or not to Bonferroni: when and how are the questions. *Bulletin of the Ecological Society of America,* **81**: 246-248.

Caroll, S.P. and Dingle, H. (1996) The biology of post-invasion events. *Biological Conservation,* **78**: 207-214.

Clement, E.J. and Foster, M.C. (1994) Alien Plants of the British Isles, London: Botanical Society of the British Isles.

Crawley, M.J. (1987) What makes a community invasible? In: *Colonization, succession and stability* (A.J. Gray, M.J. Crawley and P.J. Edwards, eds), pp. 429-453 Oxford: Blackwell Scientific Publications.

Cross, J.R. (1982) The invasion and impact of *Rhododendron ponticum* in native Irish vegetation. *Journal of Life Science of the Royal Dublin Society*, **3**: 209-220.

DeWalt, S.J. and Hamrick, J.L. (2004) Genetic variation of introduced Hawaiian and native Costa Rican populations of an invasive tropical shrub, *Clidemia hirta* (Melastomataceae). *American Journal of Botany*, **91**: 1155-1162.

Durka, W., Bossdorf, O., Prati, D. and Auge, H. (2005) Molecular evidence for multiple introduction of garlic mustard (*Alliaria petiolata*, Brassicaceae) to North America. *Molecular Ecology*, **14**: 1697-1706.

Ellstrand, N.C. and Schierenbeck, K.A. (2000) Hybridization as a stimulus for the evolution of invasiveness in plants. *Proceedings of the National Academy of Sciences of the United States of America*, **97**: 7043-7050.

Erfmeier, A. and Bruelheide, H. (2005) Invasive and native *Rhododendron ponticum* populations: Is there evidence for genotypic differences in germination and growth? *Ecography*, **28**: 417-428.

Facon, B., Jarne, P., Pointier, J.P. and David, P. (2005) Hybridization and invasiveness in the freshwater snail *Melanoides tuberculata*: hybrid vigour is more important than increase in genetic variance. *Journal of Evolutionary Biology*, **18**: 524-535.

Falconer, D.S. and Mackay, T.F.C. (1996) Introduction to Quantitative Genetics, Essex: Longman.

Gaskin, J.F. and Schaal, B.A. (2002) Hybrid *Tamarix* widespread in US invasion and undetected in native Asian range. *Proceedings of the National Academy of Sciences of the United States of America*, **99**: 11256-11259.

Hochberg, Y. (1974) Some Conservative Generalizations of the T-Method in Simultaneous Inference. *Journal of Multivariate Analysis*, **4**: 224-234.

Hollingsworth, M.L., Hollingsworth, P.M., Jenkins, G.I., Bailey, J.P. and Ferris, C. (1998) The use of molecular markers to study patterns of genotypic diversity in some invasive alien *Fallopia* spp. (Polygonaceae). *Molecular Ecology*, **7**: 1681-1691.

Houtman, R.T., Kraan, K.J. and Kromhout, H. (2004) *Mahonia aquifolium, M. repens, M. x wagneri* en hybriden. *Dendroflora*, **41**: 42-69.

Keane, R.M. and Crawley, M.J. (2002) Exotic plant invasions and the enemy release hypothesis. *Trends in Ecology & Evolution*, **17**: 164-170.

Kitajima, K., Fox, A.M., Sato, T. and Nagamatsu, D. (2006) Cultivar selection prior to introduction may increase invasiveness: evidence from *Ardisia crenata*. *Biological Invasions*, **8**: 1471-1482.

Kowarik, I. (1992) Einführung und Ausbreitung nichteinheimischer Gehölzarten in Berlin und Brandenburg. *Verhandlungen Botanischer Vereine Berlin Brandenburg*, **3**: 1-188.

Kuehn, I., and Klotz, S. (2003) The alien flora of Germany - basics from a new German database. In: *Plant invasions: ecological threats and management solutions* (L.E. Child, J.H. Brock, G. Brundu, K. Prach, P. Pysek, P.M. Wade and M. Williamson, eds), pp. 89-100 Leiden: Backhuys Publishers.

Lawrence, M.J. (1984) The genetical analysis of ecological traits. In: *Evolutionary Ecology* (B. Shorrocks, eds), pp. 27-63 Oxford: Blackwell Scientific.

Lichtenthaler, H.K. and Wellburn, A.R. (1983) Determination of total carotenoids and chlorophyll a and b of leaf extracts in different solvents. *Biochemical Society transactions*, **11**: 591-592.

Lohmeyer, W and Sukopp, H. (1992) Agriophyten in der Vegetation Mitteleuropas. *Schriftenreihe für Vegetationskunde*, Vol. 25, Bonn-Bad Godesberg: Bundesforschungsanstalt für Naturschutz und Landschaftsökologie.

Lynch, M. and Walsh, B. (1998) Genetics and Analysis of Quantitative Traits, Sunderland: Sinauer Associates, Inc. Publishers.

Mack, R.N. (2000) Cultivation fosters plant naturalization by reducing environmental stochasticity. *Biological Invasions*, **2**: 111-122.

Mandak, B., Pysek, P. and Bimova, K. (2004) History of the invasion and distribution of *Reynoutria* taxa in the Czech Republic: a hybrid spreading faster than its parents. *Preslia*, **76**: 15-64.

Maron, J.L., Vila, M., Bommarco, R., Elmendorf, S. and Beardsley, P. (2004) Rapid evolution of an invasive plant. *Ecological Monographs*, **74**: 261-280.

Milne, R.I. and Abbott, R.J. (2000) Origin and evolution of invasive naturalized material of *Rhododendron ponticum* L. in the British Isles. *Molecular Ecology*, **9**: 541-556.

Moody, M.L. and Les, D.H. (2002) Evidence of hybridity in invasive watermilfoil (*Myriophyllum*) populations. *Proceedings of the National Academy of Sciences of the United States of America*, **99**: 14867-14871.

Moran, M.D. (2003) Arguments for rejecting the sequential Bonferroni in ecological studies. *OIKOS*, **100**: 403-405.

Neuffer, B., Auge, H., Mesch, H., Amarell, U. and Brandl, R. (1999) Spread of violets in polluted pine forests: morphological and molecular evidence for the ecological importance of interspecific hybridization. *Molecular Ecology,* **8**: 365-377.

Parker, I.M., Rodriguez, J. and Loik, M.E. (2003) An evolutionary approach to understanding the biology of invasions: Local adaptation and general-purpose genotypes in the weed *Verbascum thapsus. Conservation Biology,* **17**: 59-72.

Piper, C.V. (1922) The identification of *Berberis aquifolium* and *Berberis repens. Contributions from the United States National Herbarium,* **20**: 437-451.

Preston, C.D., Telfer, M.G., Arnold, H.R., Carey, P.D., Cooper, J.M., Dines, T.D., Hill, M.O., Pearman, D.A., Roy, D.B. and Smart, S.M. (2002) The changing flora of the UK, London: DEFRA: Oxford University Press.

Reichard, S.H. and Hamilton, C.W. (1997) Predicting invasions of woody plants introduced into North America. *Conservation Biology,* **11**: 193-203.

Rieseberg, L.H., Kim, S.-C., Randell, R.A., Whitney, K.D., Gross, B.L., Lexer, C. and Clay, K. (2007) Hybridization and the colonization of novel habitats by annual sunflowers. *Genetica,* **129**: 149-165.

Roach, D.A. and Wulff, R.D. (1987) Maternal Effects in Plants. *Annual Review of Ecology and Systematics,* **18**: 209-235.

Roß, C. and Durka, W. (2006) Isolation and characterization of microsatellite markers in the invasive shrub *Mahonia aquifolium* (Berberidaceae) and their applicability in related species. *Molecular Ecology Notes,* **6**: 948-950.

Ross, C.A., Auge, H. and Durka, W. (2008) Genetic relationship among three native North-American *Mahonia* species, invasive *Mahonia* populations from Europe and commercial cultivars. *Plants Systematic and Evolution,* **275**, 219-229

Rossiter, M.C. (1996) Incidence and consequences of inherited environmental effects. *Annual Review of Ecology and Systematics,* **27**: 451-476.

Sakai, A.K., Allendorf, F.W., Holt, J.S., Lodge, D.M., Molofsky, J., With, K.A., Baughman, S., Cabin, R.J., Cohen, J.E., Ellstrand, N.C., McCauley, D.E., O'Neil, P., Parker, I.M., Thompson, J.N. and Weller, S.G. (2001) The population biology of invasive species. *Annual Review of Ecology and Systematics,* **32**: 305-332.

Sokal, R.R. and Rohlf, F.J. (1995) Biometry, New York: W.H. Freeman and Company.

Soldaat, L.L., and Auge, H. (1998) Interactions between an invasive plant, *Mahonia aquifolium,* and a native phytophagous insect, *Rhagoletis meigenii.* In: *Plant Invasions: Ecological mechanisms and human responses* (U. Starfinger, K. Edwards, I. Kowarik and M. Williamson, eds), pp. 347-360 Leiden: Backhuys Publishers.

Stace, C. (1991) New Flora of the British Isles, Cambridge: Press Syndicate of the University of Cambridge.

van de Laar, H.J. (1975) *Mahonia* en *Mahoberberis*. *Dendroflora*, **11/12**: 19-33.

Vila, M. and D'Antonio, C.M. (1998) Hybrid vigor for clonal growth in *Carpobrotus* (Aizoaceae) in coastal California. *Ecological Applications*, **8**: 1196-1205.

Vila, M., Weber, E. and D'Antonio, C.M. (2000) Conservation implications of invasion by plant hybridization. *Biological Invasions*, **2**: 207-217.

Whittemore, A.T. (1997) *Berberis*. In: *Flora of North America* pp. 276-286 New York, Oxford: Oxford University Press.

Chapter 5

Mahonia invasions in different habitats: local adaptation or general-purpose genotypes?

In cooperation with Daniela Faust and Harald Auge

5.1 Abstract

Rapid evolutionary adaptations and phenotypic plasticity have been suggested to be two important, but not mutually exclusive, mechanisms contributing to the spread of invasive species. Adaptive evolution in invasive plants has been shown to occur at large spatial scales to different climatic regions, but local adaptation at a smaller scale, e.g. to different habitats within a region, has rarely been studied. Therefore, we performed a case study on invasive *Mahonia* populations to investigate whether local adaptation may have contributed to their spread. We hypothesised that the invasion success of these populations is promoted by adaptive differentiation in response to local environmental conditions, in particular to the different soils in these habitats. To test this hypothesis, we carried out a reciprocal transplantation experiment in the field using seedlings from five *Mahonia* populations in Germany that are representative for the range of habitats invaded, and a greenhouse experiment that specifically compared the responses to the different soils of these habitats. We found no evidence for local adaptation of invasive *Mahonia* populations because seedlings from all populations responded similarly to different habitats and soils. In a second greenhouse experiment we examined genetic variation within populations, but seedlings from different maternal families did not vary in their responses to soil conditions. We therefore suggest that local adaptation of seedlings does not play a major role for the invasion success of *Mahonia* populations and that phenotypic plasticity, instead, could be an important trait in this stage of the life cycle.

5.2   Introduction

Local adaptation to environmental conditions is common among plants (Linhart and Grant 1996). For instance, studies have found adaptation to different competitors (Turkington and Harper 1979), soil conditions (Ellis and Weis 2006; Rajakaruna et al. 2003; Snaydon 1961), or climates (Joshi et al. 2001; Sawada et al. 1994; Weaver and Dirks 1984). In recent years, there has been increasing evidence that such adaptive divergence can take place at very short time-scales that are relevant for ecological processes (Conner 2003; Stockwell et al. 2003; Thompson 1998) for instance adaptation to increased atmospheric CO_2 (Wieneke et al. 2004; Ward et al. 2000), heavy metal pollution, herbicide application, or elevated pH values in the soil (Bone and Farres 2001). In addition to the importance of adaptive responses of native species to such recent environmental changes, the ability for evolutionary adjustments has been suggested to be a key feature of successful invaders (Bossdorf et al. 2005; Sakai et al. 2001; Caroll and Dingle 1996). Non-indigenous species are exposed to novel environments and thus to novel selection pressures at different spatial scales. While on a larger scale, adaptation to different climatic regions may contribute to the geographic spread of invasive species (Sexton et al. 2002; Weber and Schmid 1998; Weaver and Dirks 1984), rapid adaptation to small-scale environmental conditions may increase both the number of invaded habitats within regions and the dominance within habitats (Parker et al. 2003). However, local adaptation at smaller scales has rarely been studied in invasive species.

Adaptive divergence is a result of divergent selective forces imposed by different environments and is characterized by higher fitness of resident genotypes compared to genotypes from other habitats. Transplant experiments are useful tools to examine local adaptation (Kawecki and Ebert 2004; Bradshaw 1984), because they compare different origins of plants in their natural environments, and fitness of local and foreign populations can be directly compared. Despite their usefulness, only few reciprocal transplant experiments have been applied to invasive species (Parker et al. 2003; but see Rice and Mack 1991). Since the response of a trait to selection is proportional to its heritable variation (Lynch and Walsh 1998), local adaptation is constrained by the amount of genetic variation present in the population (Kawecki and

Ebert 2004). In invasive species, however, genetic bottlenecks are expected to be common which may impede evolutionary adjustments (Allendorf and Lundquist 2003). Repeated introductions from different source populations (Bossdorf et al. 2005) and interspecific hybridisation (Ellstrand and Schierenbeck 2000) may counteract a genetic bottleneck and facilitate the invasion of different habitats. Apart from local adaptation, general-purpose genotypes (Baker 1965), i.e. genotypes with high phenotypic plasticity, may be another, but not mutually exclusive, explanation for the successful spread of invasive species (e.g. Parker et al. 2003). Since phenotypic plasticity has a genetic basis and is subjected to selection (Schlichting 1986), it can be suggested that invasive populations have evolved greater plasticity than conspecific populations in the native range (Richards et al. 2006; Bossdorf et al. 2005). Phenotypic plasticity may than enhance niche breadth and confer a fitness advantage for successful invaders in a broad range of environments (Richards et al. 2006). The invasion ability of a foreign species into different habitats may therefore be based on general-purpose genotypes and/or on local adapted genotypes.

In this chapter, we report on a series of reciprocal transplant experiments that tested for local adaptation of invasive *Mahonia* populations to different habitats in Germany. We supposed that local adaptation rather than phenotypic plasticity is important for the invasion of *Mahonia* into different habitats, because high genetic variation exist in invasive populations, that may have fostered adaptation. To demonstrate local adaptation in transplant experiments, genotypes growing at their home site (i.e. local genotypes) should be superior over genotypes transplanted from a foreign site ('local vs. foreign' criterion: (Kawecki and Ebert 2004)).

Mahonia aquifolium (PURSH) NUTT. (Berberidaceae) is a shrub native in western North America that was introduced to Europe for ornamental purpose about 180 years ago (Hayne 1822, cited in Kowarik 1992). The descendants of the cultivated plants have become invasive in Central Europe and occur across a wide range of habitats from mixed forests and dry scrub vegetation on calcareous soils to pine forests on sandy soils or even wall breaks (Auge and Brandl 1997; Kowarik 1992; Lohmeyer and Sukopp 1992). Since there is evidence of a hybrid origin (Ahrendt 1961) we will use the term invasive *Mahonia* populations instead of *M. aquifolium*.

In our study, we focused on the seedling stage of *Mahonia*. Since survival rate is usually poor among seedlings, adaptation can be crucial for survival in this life stage and fitness differences among genotypes can be detected (Primack and Kang 1989). Seedling mortality is also very high in *Mahonia* populations (Auge and Brandl 1997), and we thus expect selection to affect especially seedlings. Seedling recruitment contributes to the colonisation of new sites as well as to local population growth in *Mahonia* (Auge and Brandl 1997). In addition, seedlings are more manageable than adults, particular in woody species. Specifically, we tested the hypotheses that (1) seedlings of invasive *Mahonia* populations perform better at their home site and in their home soil than seedlings transplanted from foreign sites, and (2) maternal families within *Mahonia* populations differ in their response to soil conditions, indicating genetic variation as a precondition for evolutionary adjustments.

5.3   Material and Methods

5.3.1   Study species

Oregon grape, *Mahonia aquifolium* (PURSH) NUTT. (Berberidaceae), is a fleshy-fruited evergreen shrub native to western North America. The species was introduced for ornamental purposes to central Europe in 1822 (Hayne 1822, cited in Kowarik 1992). Cultivated *M. aquifolium* plants are morphologically highly variable (Ahrendt 1961), drought-resistant, and grow in a wide range of light and soil conditions (Houtman et al. 2004). The first spontaneous occurrence outside gardens was observed 38 years after the date of introduction (Kowarik 1992). Today, the shrub is a widespread invader of anthropogenic as well as natural vegetation in Central Europe (Kowarik 1992; Lohmeyer and Sukopp 1992). Cultivated *M. aquifolium* were hybridised with the related, North American species *Mahonia repens* (LINDL.) G.DON and *M. pinnata* (LAG.) FEDDE. (Ahrendt 1961), resulting in many cultivars (van de Laar 1975). It is therefore likely that invasive populations consist largely of hybrids. Invasive *Mahonia* populations reproduce sexually by seedlings, and clonally by stolons, root sprouts and stem layers (Auge 1997). However, sexually reproduction seems to play a major role to regional as well as local spread (Auge 1997). *Mahonia* individuals flowers usually as from the third year of growth (C.A. Ross, personal observation) from April to June and produce multitudes of berries from August to October, that may stick to the sprout till winter (Zeitlhöfler 2002). *Mahonia* is supposed to be an outbreeding species like the whole genus (Burd 1994), in particular due to self incompatibility since self-pollination does rarely result in fruit production (Monzingo 1987, H. Auge unpublished data). Invasive *Mahonia* populations show great variation in quantitative traits (Auge et al. 1997; Ahrendt 1961) which has a large genetic component (H. Auge, unpublished data).

5.3.2   Study sites and characterisation of habitat conditions

We chose five invasive *Mahonia* populations from different regions of the species' distribution range in Germany (Figure 1) that were well established and differed in habitat conditions (Table 1). The spatial distance between the five populations ranged from 102 km to 526 km. To verify that habitat conditions differed between the five sites we measured soil moisture, soil chemical properties and relative irradiance in each. Soil moisture and relative irradiance were measured three times in the summer of 2004, at intervals of approximately four weeks. We used time domain reflectometry (TDR Multimeter, Easy Test Ltd., Lublin, Poland) to determine moisture within the upper 10 cm of the soil. At each site and at each date, we took the measurements at eight random locations separated by at least 10 m. At each, we took three soil moisture readings at distances of 10 cm that were averaged thereafter. Furthermore, we determined relative irradiance as percent of photosynthetic active radiation (PhAR) penetrating the canopy of trees and reaching *Mahonia* shrubs using a Li-191SA Line Quantum Sensor (LI-COR, Nebraska, USA). At each site and date, we took PhAR measurements at ten random locations on the level of *Mahonia* shoot tips. For soil chemical analyses, we sampled the upper soil layer at six locations within each site. We determined pH values in 0.1 n KCl solution using a Calimatic pH meter Typ 765 (Knick elektronische Messgeräte GmbH & Co, Berlin, Germany), and measured carbonate content using a Scheibler calcimeter (Schlichting et al. 1995). Further, we determined C and N concentrations by dry combustion and subsequent gas analysis using an Element Analyser Vario EL (Elementar Analysesysteme GmbH, Hanau, Germany). For analyses of plant available K and P concentrations, we used double-lactate extraction and subsequent x-ray fluorescence analysis of the extract. Total concentrations of macronutrients (Mg, Ca, P, K), trace elements (Fe, Al), and Si was also determined by x-ray fluorescence analysis. We applied the same analyses to the soils used in the greenhouse experiments (see below).

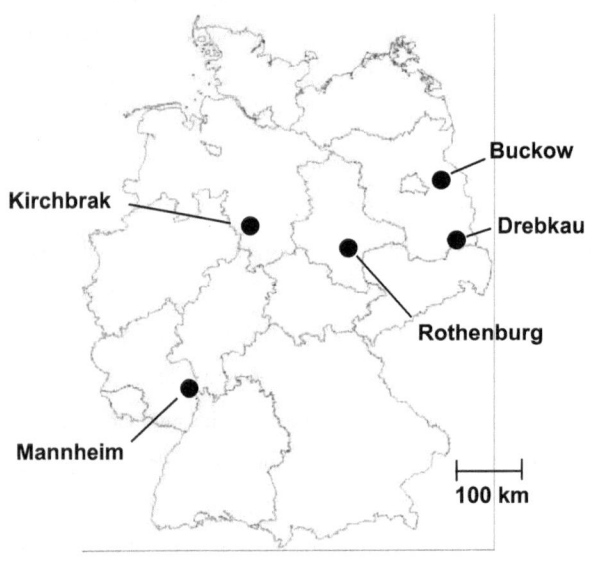

Figure 1: Geographic locations of the invasive _Mahonia_ populations investigated in this study. The map shows Germany and, for orientation, the boundaries of the federal states.

Table 1: Geographic location, vegetation type and site conditions of the five invasive _Mahonia_ populations

Location	Geographic coordinates	Vegetation	Soil type	Geology	Mean annual temperature [°C]	Mean annual rainfall [mm]
Buckow (Bu)	52°30'N 14°08' E	hardwood forest	Cambisol	sand	8.3	531.0
Drebkau (Dr)	51°39'N 14°13'E	pine forest	Cambisol	sand	8.9	563.0
Kirchbrak (Ki)	51°58'N 9°35'E	mixed forest	Leptosol	shell limestone	9.2	720.1
Mannheim (Ma)	49°29'N 8°28'E	hardwood forest	Cambisol	calcareous eolian sand	10.2	667.6
Rothen- burg (Ro)	51°38'N 11°45'E	dry scrub	Cambisol/ Leptosol	carbonic sand stone	9.2	469.2

5.3.3   Rearing of seedlings

In August 2003, we collected seeds from ten individuals along a transect in each of the five populations. We collected 30-50 berries from each individual. The offspring of each maternal plant represents at least half-sibs and will therefore in the following be referred to as a seed family. In December 2003, we sowed the seeds in plastic trays containing a mixture of 50 % rearing compost (Composana; COMPO GmbH, Münster, Germany) and 50 % sand, and stored them at 5°C in a refrigerator. After 16 weeks of stratification, we transferred the seed trays to a climate chamber with a 14 h/10 h day/night cycle at 15°C/10°C. Three weeks later we raised the day temperature to 20°C to facilitate germination. After further 2 weeks, we planted the seedlings separately in pots with 50 % of a standard potting soil (Fruhstorfer Typ P, florimaris Humus- und Erdenwerk GmbH & Co.Kg, Wangerland, Germany) and 50 % sand, and placed them in a greenhouse with natural light conditions and a day/night cycle of 14 h/10 h and 25°C/15°C until the start of the experiments.

5.3.4   Field experiment

To test for a general pattern of local adaptation in invasive *Mahonia* populations we set up a reciprocal transplant experiment in the field. In June 2004, we planted ten seedlings from each of the five populations at each of the five study sites. Seedlings from each population were thus planted at their home site and at all other sites. Each of the ten seedlings planted at one site originated from a different seed family, with the same set of seed families per population used in each site. At each site, seedlings were planted in five blocks, with two seedlings of each population in each block. The seedlings were planted carefully to reduce disturbance of soil and neighbouring vegetation, and watered once after planting. We measured survival after 14 months in August 2005, harvested the surviving seedlings and measured their aboveground dry mass.

5.3.5   Greenhouse experiments

In order to test for local adaptation to soil conditions, we carried out a first greenhouse experiment that compared the growth responses of seedlings from the five *Mahonia* populations to the soils from each site. For this purpose, we took soil samples from eight random locations at each of the five study sites and prepared a soil mixture of each site after removing big stones and plant remnants. To verify whether these mixtures mirrored soil conditions in the field we carried out chemical analyses of the soil mixtures in the same way as described above. Each soil mixture was filled into 60 250-ml plastic pots. We planted 12 seedlings per population, originating from different seed families and reared as described above, in each soil mixture. We placed the pots on two tables in each of two greenhouse chambers (three seedlings of each population x soil mixture combination on each table). Because the habitats from which soil was taken differed in soil moisture, and in order to avoid inappropriate soil-water combinations in the greenhouse, we established two soil moisture levels by watering one greenhouse table in each chamber twice a week ("low soil moisture") and the other one three times a week ("high soil moisture"). These two treatments corresponded to soil moisture of 9.2 ± 1.1 vol % and 11.2 ± 1.7 vol %, respectively (means $\pm$ standard errors across the five soils, measured on 6 days immediately before watering). The pots were randomly placed on each table and re-arranged every week. Every four weeks, we redistributed the pots between the two climate chambers. This allowed us to ignore the effects of table and greenhouse chamber in the subsequent data analyses. Thus, the experiment was a three-factorial with six replicates of each population x soil x soil moisture combination. After three months of growth in the greenhouse (14 h/10 h day/night, 25°C/15°C), we harvested the seedlings and measured their above-ground dry mass. As indication of developmental stage, we determined whether the seedlings still had entire (primary) or already pinnate (mature) leaves. Pinnate leaves can be used to distinguish the sapling stage from the seedling stage in this species (Auge and Brandl 1997).

In a second greenhouse experiment, we tested for genetic variation within populations for the five different soil conditions. Since each seed family consisted of at least of half-sibs, the comparison among seed families gave an estimate of genetic variation in plant traits (Falconer and Mackay 1996; Lawrence 1984). To measure within-population variation in the response to soil conditions, we studied the reaction norms of four randomly selected seed families within each of three invasive *Mahonia* populations (Dr, Ki, Ro). Seedlings were planted individually in 250 ml plastic pots filled with one of the five soils, with three replicates of each seed family x soil combination resulting in a total of 180 pots. The pots were placed on a greenhouse table and watered three times a week. Day length and temperatures were the same as described for the first greenhouse experiment. We harvested aboveground parts of the seedlings after three months of growth, and determined their dry mass and the presence of pinnate leaves.

5.3.6   Statistical analyses

For all statistical analyses we used SAS version 8.2 (SAS Institute, Cary, NC, USA). To compare relative irradiance and soil moisture between the five sites, we carried out analysis of variance (ANOVA) with site and sampling date nested within site as fixed factors (SAS PROC GLM). To meet the requirements of the ANOVA we applied arcsine square-root transformation to the data. To compare soil chemical properties between the five sites, we carried out univariate analyses of variance for each variable followed by sequential Bonferroni tests (Dunn-Sydak method, Sokal and Rohlf 1995), and a multivariate analysis of variance (MANOVA) with 13 soil variables as dependent variables (PROC GLM). We abandoned C concentrations and included only N concentrations and C/N ratios in the MANOVA to avoid redundant data.

We analysed biomass data from the field experiment using mixed model analysis of variance (ANOVA) (PROC MIXED). We regarded population and site as fixed factors since the sites and populations were deliberately chosen to maximize differences in habitat conditions. Block within sites was considered as a random factor. Since mortality of seedlings was high and the data were therefore unbalanced, error degrees of freedom for the test of fixed effects were computed using the Satterthwaite approximation. Wald Z-test was applied to the variance estimates attributable to random effects in the model. The probability of survival was analysed in a similar way except that we used a generalised linear model with binomial error distribution and logit link function (PROC GENMOD). We calculated quasi-F-tests for all effects and tested the site effect against the variation among blocks within sites.

For statistical analysis of biomass data from the first greenhouse experiment, we carried out a mixed model ANOVA using population, soil, and soil moisture as fixed factors (PROC MIXED). We analysed the probability to produce pinnate leaves using a generalised linear model, again with population, soil and moisture as fixed factors (PROC GENMOD). The population x site and population x soil interactions of the ANOVA models provide evidence whether populations respond differently to the five soils or sites, and is therefore a test to detect local adaptation.

Seedling biomass in the second greenhouse experiment was also analysed with a mixed model ANOVA (PROC MIXED). Population and soil origin were treated as fixed factors, while seed family within population was considered as a random factor. We, therefore, tested the population effect against the seed family within population x soil effect and the soil effect as well as the population x soil interaction effect against seed family within population x soil. Random effects were tested with Wald Z-statistics. We analysed the probability to produce pinnate leaves again using a generalised linear model with soil and population as fixed factors and seed family within population as random factor (PROC GENMOD). We consider the seed family main effect and the seed family x soil interaction effect as evidence for genetic variation within populations.

Prior to all analyses, we logarithmically transformed biomass to approach normal distribution and homoscedasticity.

5.4   Results

5.4.1   Habitat conditions

There were considerable differences in irradiance and soil moisture between the five study sites (Table 2), with irradiance values between 12 % and 80 %, and soil moisture values of 6 % to 14 %. In addition, MANOVA revealed large variation in soil chemical properties between the five sites (Wilk's Lambda = 1.1 x 10^{-6}, $F = 33.84$, $p < 0.0001$; univariate statistics are given in Table 2).

Table 2: Mean values (± standard deviation) and univariate comparisons of habitat conditions (degrees of freedom for F statistics: 4, 135 for relative irradiance; 4, 105 for soil moisture, 4, 25 for soil chemical properties; * $p < 0.05$; ** $p < 0.01$; *** $p < 0.001$; sequential Bonferroni tests applied to soil chemical properties) ([1] total content, [2] plant available content).

Site	Bu	Dr	Ki	Ma	Ro	F value
Relative irradiance [%]	30.5 ± 17.9	17.0 ± 10.0	11.5 ± 7.5	22.1 ± 27.4	80.1 ± 30.7	52.52 ***
Soil moisture [Vol %]	6.36 ± 2.27	7.23 ± 2.00	11.2 ± 2.5	5.63 ± 1.82	13.7 ± 4.3	39.05 ***
pH value[1]	4.70 ± 0.80	4.95 ± 0.17	5.94 ± 0.80	6.54 ± 1.02	7.62 ± 0.30	21.47 ***
N content [%][1]	0.146 ± 0.074	0.235 ± 0.076	0.502 ± 0.130	0.293 ± 0.099	0.245 ± 0.144	9.02 ***
C/N ratio[1]	16.0 ± 2.0	20.7 ± 1.6	11.7 ± 1.0	18.0 ± 3.0	20.7 ± 8.0	8.08 ***
$CaCO_3$ content [%][1]	0 ± 0	0 ± 0	0.913 ± 0.638	0.793 ± 0.787	5.95 ± 2.20	37.50 ***
MgO content [%][1]	0.141 ± 0.014	0.154 ± 0.017	2.03 ± 0.38	0.359 ± 0.053	0.926 ± 0.131	22.89 ***
Al_2O_3 content [%][1]	2.72 ± 0.10	2.45 ± 0.20	13.2 ± 1.3	4.54 ± 0.37	7.52 ± 0.33	579.91 ***
SiO_2 content [%][1]	90.1 ± 1.9	84.0 ± 3.37	54.1 ± 6.2	80.8 ± 2.9	70.7 ± 3.8	79.13 ***
P_2O_5 content [ppm][1]	593 ± 101	670 ± 95	1486 ± 352	688 ± 94	1075 ± 183	22.89 ***
K_2O content [%][1]	0.988 ± 0.044	0.833 ± 0.033	3.53 ± 0.31	1.59 ± 0.08	2.25 ± 0.10	655.30 ***
CaO content [%][1]	0.338 ± 0.082	0.410 ± 0.073	1.48 ± 0.73	1.26 ± 0.50	3.80 ± 1.02	42.70 ***
Fe_2O_3 content [%][1]	0.501 ± 0.056	1.17 ± 0.23	5.38 ± 0.75	0.879 ± 0.150	2.82 ± 0.18	256.34 ***
K content [ppm][2]	79.4 ± 11.7	97.9 ± 10.8	317 ± 55	1379 ± 27	238 ± 82	46.14 ***
P content [ppm][2]	27.8 ± 4.5	14.7 ± 1.5	17.9 ± 6.2	16.3 ± 2.1	15.3 ± 8.4	5.18 **

In the PCA, the first two axes explained 61 % and 22 % of the variation in soil chemical properties, respectively. The first PCA axis was positively correlated with nutrient contents, and negatively correlated with silicon content. The second PCA axis correlated positively with pH-value and carbonate, and negatively with phosphorus content. While samples from the same site clustered together fairly well, the sites were clearly separated by their scores on the first two PCA axes (Figure 2). Thus, all five study sites differed in the habitat conditions experienced by the local *Mahonia* populations.

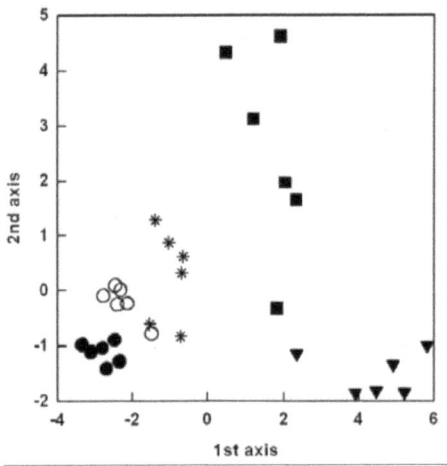

Figure 2: Results of the principle component analysis of soil chemical properties of the five sites (●=Bu, ○=Dr, ▼=Ki, ✳=Ma, ■=Ro). The first and second principle components together explain 83 % of the variance.

5.4.2   Reciprocal transplantation in the field

By the time of the harvest, 47 % of originally planted seedlings were still alive. While survival rate did not differ significantly between populations or study sites, there were significant block effects indicating small-scale spatial variation in seedling survival within sites (Table 3). Although the plants were still rather small at the time of harvesting, their average biomass was increased by 176 % when compared to their size at the time of planting 14 months earlier. Seedling biomass differed significantly between sites but there was no difference between populations. On the "best" site (Ma),

Table 3: Results of mixed-model ANOVAs for the effects of population and site on aboveground dry biomass and survival probability of *Mahonia* seedlings in a reciprocal transplantation experiment in the field (* $p < 0.05$; ** $p < 0.01$; *** $p < 0.001$; variance estimates for random effects have 1 df)

Source of variation	Dry mass		Survival	
Fixed effects	Df	F values	df	F values
Population	4, 94	1.85	4, 203	1.23
Site	4, 94	3.14*	4, 20	1.19
Site x population	16, 94	0.92	16, 203	0.73
Random effects	Variance estimate	Z values	df	F values
Block within site	0.0000	---	20, 203	5.11***
Residuals	0.8570	6.86***	203	---

Table 4: Results of mixed-model ANOVAs for the effects of population, soil origin and soil moisture on growth of *Mahonia* seedlings in a greenhouse experiment (* $p < 0.05$; ** $p < 0.01$; *** $p < 0.001$).

Source of variation	Dry mass		Probability of pinnate leaves	
Fixed effects	Df	F values	Df	F values
Population	4, 244	1.11	4, 244	1.13
Soil	4, 244	80.97***	4, 244	18.94***
Soil moisture	1, 244	4.98*	1, 244	0.52
Population x soil	16, 244	1.20	16, 244	0.72
Population x soil moisture	4, 244	0.64	4, 244	0.32
Soil x soil moisture	4, 244	1.38	4, 244	1.09
Population x soil x soil moisture	16, 244	0.77	16, 244	0.96

average seedling biomass was almost four times higher (153 ± 28 mg, mean ± s.e.) than on the "worst" site (Ro, 42 ± 20 mg).

Populations did not respond differently to these site conditions as indicated by a lack of significant population x site interactions for survival and biomass (Figure 3). Seedling biomass in the field experiment did not correlate with average seedling biomass on the respective soils in the greenhouse experiment ($r = 0.40$, $p = 0.50$).

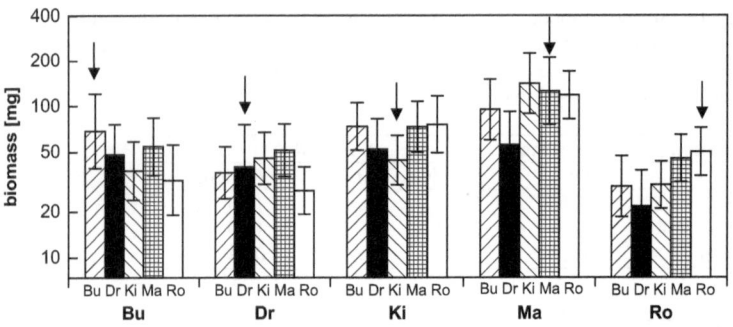

Figure 3: Mean aboveground dry mass of seedlings from the five invasive *Mahonia* populations in a reciprocal transplantation experiment in the field (means ± standard error). Each bar corresponds to one population, groups of bars represent the different sites, and the arrows highlight seedlings grown at their 'home' site. The population x site interaction is not significant (see table 3).

5.4.3   Greenhouse experiments

At the end of the first greenhouse experiment, *Mahonia* seedlings had an average biomass of 241 mg, and 43 % of the seedlings had produced at least one pinnate leaf indicating an advanced ontogenetic stage. The presence of compound, pinnate leaves can be used to distinguish the sapling stage from the seedling stage in this species, because *Mahonia* seedling have only entire (primary) leaves (Auge and Brandl 1997). There were no differences between the five populations, but soil strongly affected seedling biomass and the probability to produce pinnate leaves. For instance, seedlings had an average biomass of 126 ± 6 mg and 23 ± 6 % of them produced pinnate leaves in the "worst" soil (Ro), whereas seedling biomass and the probability to produce pinnate leaves were 446 ± 16 mg and 86 ± 6 %, respectively, in the "best" soil (Ki). In contrast to the strong effect of soil, the watering regime had only a weak influence as plant biomass was about 18 % larger at high soil moisture compared to the low soil moisture treatment. Populations responded similarly to the different soils, i.e. the population x soil interaction was not significant (Figure 4).

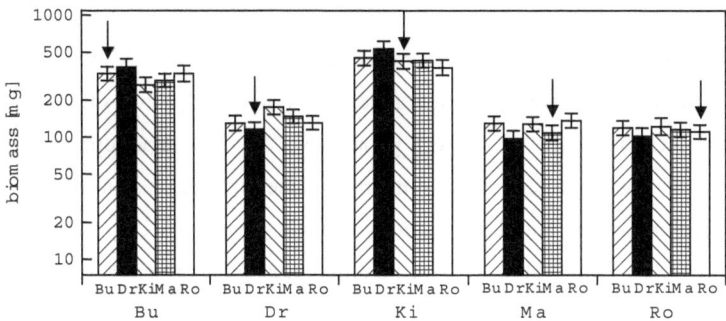

Figure 4: Mean aboveground dry mass of seedlings from five invasive *Mahonia* populations grown reciprocally in five different soils in the greenhouse (means ± standard error). Each bar corresponds to one population, groups of bars represent the different soils, and the arrows highlight seedlings grown in their 'home' soil. The population x soil interaction is not significant (see table 4).

At the end of the second greenhouse experiment, mean seedling biomass was 237 mg and 39 % of seedlings had produced pinnate leaves. Whereas populations did not differ in biomass, they varied significantly in their probability to build pinnate leaves. Again, there was a strong effect of soil origin on biomass and probability to build pinnate leaves. There was, however, neither a significant variation among seed families within populations nor a significant seed family x soil interaction. Thus, reaction norms to different soils were similar across seed families, indicating a lack of genetic variation for plasticity to soil conditions.

Table 5: Results of mixed-model ANOVAs for the effects of population, seed family and soil on growth of *Mahonia* seedlings in a greenhouse experiment (* $p < 0.05$; ** $p < 0.01$; *** $p < 0.001$; variance estimates for random effects have 1 df).

Source of variation	Dry mass		Probability of pinnate leaves	
Fixed effects	Df	*F* values	Df	*F* values
Population	2, 9	0.73	2, 9	11.06**
Soil	4, 36	48.49***	4, 36	17.96***
Population x soil	8, 36	1.99	8, 36	2.25*
Random effects	Variance estimate	*Z* values	Df	*F* values
Seed family (Population)	0.0109	0.89	9, 119	0.79
Soil x seed family (Population)	0.0000	---	36, 119	1.13
Residuals	0.2208	8.80***	119	---

5.5   Discussion

With respect to our initial hypotheses, we can summarise our results as follows: (1) There were great differences among the five field sites in terms of growth of *Mahonia* seedlings. However, seedlings of resident populations did not outperform foreign seedlings, but all seedlings responded similarly to the different habitats. In the greenhouse, seedling growth and the probability to build pinnate leaves was strongly affected by the different soils, but there were no differences in the responses of the five *Mahonia* populations. (2) We found no evidence for genetic variation within populations in the response to soil conditions.

An important prerequisite for local adaptation is strong divergent selection imposed by different environments (Kawecki and Ebert 2004). It is reasonable to assume strong divergent selection between different climatic regions and therefore not surprising that several other studies on invasive plants found evidence for local adaptation to climatic factors, such as precipitation (Rice and Mack 1991), temperature (Sexton et al. 2002), and the length of the growing season (Kollmann and Bañuelos 2004; Weber and Schmid 1998). However, because of the sedentary nature of plants, divergent selection among plant populations may also occur on small spatial scales (Fritsche and Kaltz 2000; Bradshaw 1984). In addition, genetically effective neighbourhood area is usually small in plant populations (Levin 1988), and small-scale genetic differentiation can take place despite considerable levels of gene flow if environmental gradients are strong (Linhart and Grant 1996). To investigate the role of local adaptation for *Mahonia* invasions, we specifically chose five study sites which are all located in the moist, warm temperate climate-region in Central Europe (Kottek et al. 2006). The temperatures are moderate and rain falls all the year round with a slight maximum in summer. Despite the location within the same climatic region, the five sited showed great differences in vegetation composition, and our measurements of environmental variables confirmed that the sites differed strongly in abiotic habitat conditions (Table 2). These differences in habitat quality were reflected in the large variation in mean performance of *Mahonia* seedlings between the five sites. We assume that these environmental differences should have been sufficient for divergent selection to occur among our *Mahonia* populations. However, in contrast to our expectation, we found no evidence for local adaptation in the reciprocal transplant experiment in the field or in the greenhouse experiment.

Although reciprocal transplant experiments are an important tool to detect local adaptation, only few previous studies used this approach with invasive plants. For instance, Rice & Mack (1991) reciprocally sowed seeds of invasive *Bromus tectorum* at seven sites along a climatic gradient in North America and found local adaptation to the most extreme habitats. In contrast to invasive plants, there are many reciprocal transplant experiments that demonstrated local adaptation in native plant species (e.g. Bischoff et al. 2006; Joshi et al. 2001; Waser and Price 1985; Turkington and Harper 1979). In fact, there are only few examples of transplant experiments that did *not* find any evidence for local adaptation in plants (e.g. Bischoff et al. 2006).

In this study we hypothesised that, because climatic conditions were very similar, soil properties should be the most important factors responsible for differences in habitat quality and selection regimes. However, there was no correlation between seedling biomass on the five soils (measured in the greenhouse) and mean seedling biomass at the field sites. Hence, other environmental factors, such as light, water, or competitors, or interactions among them, must be more important aspects of habitat quality for *Mahonia* seedlings.

We found large differences in seedling biomass between the five soils in our greenhouse experiment, but again, there was no evidence for local adaptation of *Mahonia* populations to the soil conditions of their home site. Although we had only 10 and 12 replicates of each population x environment combination in the field or in the greenhouse experiment, respectively, we feel confident that the lack of evidence for local adaptation is not an artefact of low statistical power. There was not the slightest tendency that seedlings performed among the best in their home environment. Rather, our results show that they performed, on average, often much worse than seedlings from foreign sites. In contrast to our results, studies on native plant species have frequently found local adaptation to soil conditions, e.g. in the grass *Agrostis tenuis* (Bradshaw 1960) and several Aizoaceae species (Ellis and Weis 2006). Still, these results were based on reciprocal transplant experiments, and it is therefore probable that other factors beside soil properties also affected plant responses. To disentangle the effects of soil conditions and other environmental factors in our study system, we conducted not only a transplantation experiment in the field, but also a transplantation experiment in the greenhouse using soils from the different field sites. In addition, we manipulated the water regime in the greenhouse experiment to avoid inappropriate soil-water combinations. However, water regime did neither influence the soil effect nor the population response to soil.

In the following, we will discuss possible reasons for this lack of local adaptation in invasive *Mahonia* populations. The response of plant populations to natural selection is proportional to the genetic variation present. However, genetic variation in exotic species is often reduced by genetic bottlenecks (Allendorf and Lundquist 2003). In the case of *Mahonia*, multiple introductions and interspecific hybridisation have counteracted this bottleneck (Ross et al. submitted). Here, we used variation among maternal families as a measure of within-population genetic variation. The seed families were produced by open pollination in the field rather than by controlled mating under identical conditions (cf. Falconer and Mackay 1996; Lawrence 1984). Phenotypic variation among them may therefore have a genetic and a maternal environmental component and thus overestimate genetic variation. Still, we did not find any genetic variation within *Mahonia* populations in response to habitat or soil conditions. We can not rule out therefore that a lack of genetic variation may have prevented genetic adaptation to local environments.

Local adaptation can be counteracted by strong gene flow from other populations (Kawecki and Ebert 2004). Although our sites were more than 100 km distant from each other and pollen or seed dispersal among them should be negligible, gene flow from wild-growing populations or planted individuals in the neighbourhood may have affected our study populations. Another reason for the lack of local adaptation may be the relatively short time since the introduction of *Mahonia* in 1822. The first spontaneous occurrence of *Mahonia* outside gardens was recorded in 1860 (Kowarik 1992), and the populations we chose are probably not older than 50-80 years (cf. Auge 1997). This time-span may be too short for measurable responses to natural selection in a long-lived clonal species such as *Mahonia* (Auge and Brandl 1997). Moreover, the differences in selection regimes imposed by our study sites are probably not as drastic as in many well-known studies of local adaptation, which often studied heavy metal or herbicide influences (Bone and Farres 2001).

Finally, although we found no adaptation in the seedling stage, local adaptation could play a role in later life stages of *Mahonia*. Since selection is assumed to be particularly strong during the seedling stage (Hufford and Hamrick 2003; Primack and Kang 1989) and because seedling recruitment is important for the invasion dynamics of *Mahonia* (Auge and Brandl 1997) we focused on seedlings. However, natural selection can act during all life stages of a plant, e.g. through adult survival, pollination success, seedling germination, or clonal growth (McGraw and Antonovics 1983). Irrespective of possible adaptations in later life cycle stages, invasion of the different habitats by *Mahonia* populations must involve successful seedling recruitment. Considering the lack of local adaptation found in this study, phenotypic plasticity in the seedling stage may have contributed to the success of *Mahonia*. Our results show that *Mahonia* seedlings can sustain unfavorable conditions, but show increased growth in better environments, probably due to plasticity in physiological or morphological traits. This pattern of plasticity combines both robustness and opportunism, and is consistent with the Jack-and-master scenario of Richards et al. (2006). Because *Mahonia* seedlings are capable to establish under a wide range of environmental conditions, they show characteristics of general-purpose genotypes. We therefore suggest that the invasion success of *Mahonia* in Central Europe can, at least in part, be explained by phenotypic plasticity during early stages of the life cycle.

References

Ahrendt, L.W.A. (1961) *Berberis* and *Mahonia*. A taxonomic revision. *Journal of the Linnean Society of London, Botany,* **57**: 1-410.

Allendorf, F.W. and Lundquist, L.L. (2003) Introduction: Population Biology, Evolution, and Control of Invasive Species. *Conservation Biology,* **17**: 24-30.

Auge, H. (1997) Biologische Invasionen: Das Beispiel *Mahonia aquifolium*. In: *Regeneration und nachhaltige Landnutzung - Konzepte für belastete Regionen* (R. Feldmann, K. Henle, H. Auge, J. Flachowsky, S. Klotz and R. Kroenert, eds), pp. 124-129 Berlin: Springer Verlag.

Auge, H. and Brandl, R. (1997) Seedling recruitment in the invasive clonal shrub, *Mahonia aquifolium* Pursh (Nutt.). *Oecologia,* **110**: 205-211.

Auge, H., Brandl, R. and Fussy, M. (1997) Phenotypic variation, herbivory and fungal infection in the clonal shrub *Mahonia aquifolium* (Berberidaceae). *Mitteldeutsche Gesellschaft für Allgemeine und Angewandte Entomology,* **11**: 747-750.

Baker, H.G. (1965) Characteristics and Modes of Origin of Weeds. In: *The Genetics of Colonizing Species: Proceedings of the First International Union of Biological Sciences Symposia on General Biology* (H.G. Baker and G.L. Stebbins, eds), pp. 147-168 New York: Academic Press Inc.

Bischoff, A., Cremieux, L., Smilauerova, M., Lawson, C.S., Mortimer, S.R., Dolezal, J., Lanta, V., Edwards, A.R., Brook, A.J., Macel, M., Leps, J., Steinger, T. and Müller-Schärer, H. (2006) Detecting local adaptation in widespread grassland species - the importance of scale and local plant community. *Journal of Ecology,* **94**: 1130-1142.

Bone, E. and Farres, A. (2001) Trends and rates of microevolution in plants. *Genetica,* **112-113**: 165-182.

Bossdorf, O., Auge, H., Lafuma, L., Rogers, W.E., Siemann, E. and Prati, D. (2005) Phenotypic and genetic differentiation between native and introduced plant populations. *Oecologia,* **144**: 1-11.

Bradshaw, A.D. (1960) Population differentiation in *Agrostis tenuis* SIBTH. *New Phytologist,* **59**: 92-103.

Bradshaw, A.D. (1984) Ecological significance of genetic variation between populations. In: *Perspectives on Plant Population Ecology* (R. Dirzo and J. Sarukhán, eds), pp. 213-228 Sunderland: Sinauer Associates Inc.

Burd, M. (1994) Bateman principle and plant reproduction - the role of pollen limitation in fruit and seed set. *Botanical Review,* **60**: 83-139.

Caroll, S.P. and Dingle, H. (1996) The biology of post-invasion events. *Biological Conservation,* **78**: 207-214.

Conner, J.K. (2003) Artificial Selection: A powerful tool for ecologists. *Ecology,* **84**: 1650-1660.

Ellis, A.G. and Weis, A.E. (2006) Coexistence and differentiation of 'flowering stones': the role of local adaptation to soil microenvironment. *Journal of Ecology,* **94**: 322-335.

Ellstrand, N.C. and Schierenbeck, K.A. (2000) Hybridization as a stimulus for the evolution of invasiveness in plants. *Proceedings of the National Academy of Sciences of the United States of America,* **97**: 7043-7050.

Falconer, D.S. and Mackay, T.F.C. (1996) Introduction to Quantitative Genetics, Essex: Longman.

Fritsche, F. and Kaltz, O. (2000) Is the *Prunella* (Lamiaceae) hybrid zone structured by an environmental gradient? Evidence from a reciprocal transplant experiment. *American Journal of Botany*, **87**: 995-1003.

Houtman, R.T., Kraan, K.J. and Kromhout, H. (2004) *Mahonia aquifolium*, *M. repens*, *M. x wagneri* en hybriden. *Dendroflora*, **41**: 42-69.

Hufford, K.M. and Hamrick, J.L. (2003) Viability selection at three early life stages of the tropical tree, *Platypodium elegans* (Fabaceae, Papilionoideae). *Evolution*, **57**: 518-526.

Joshi, J., Schmid, B., Caldeira, M.C., Dimitrakopoulos, P.G., Good, J., Harris, R., Hector, A., Huss-Danell, K., Jumpponen, A., Minns, A., Mulder, C.P.H., Pereira, J.S., Prinz, A., Scherer-Lorenzen, M., Siamantziouras, A.-S.D., Terry, A.C., Troumbis, A.Y. and Lawton, J.H. (2001) Local adaptation enhances performance of common plant species. *Ecology Letters*, **4**: 536-544.

Kawecki, T.J. and Ebert, D. (2004) Conceptual issues in local adaptation. *Ecology Letters*, **7**: 1225-1241.

Kollmann, J. and Bañuelos, M.J. (2004) Latitudinal trends in growth and phenology of the invasive alien plant *Impatiens glandulifera* (Balsaminaceae). *Diversity and Distributions*, **10**: 377-385.

Kottek, M., Grieser, J., Beck, C., Rudolf, B. and Rubel, F. (2006) World Map of the Köppen-Geiger climate classification updated. *Meteorologische Zeitschrift*, **15**: 259-263.

Kowarik, I. (1992) Einführung und Ausbreitung nichteinheimischer Gehölzarten in Berlin und Brandenburg. *Verhandlungen Botanischer Vereine Berlin Brandenburg*, **3**: 1-188.

Lawrence, M.J. (1984) The genetical analysis of ecological traits. In: *Evolutionary Ecology* (B. Shorrocks, eds), pp. 27-63 Oxford: Blackwell Scientific.

Levin, D.A. (1988) Local differentiation and the breeding structure of plant populations. In: *Plant Evolutionary Biology* (L.D. Gottlieb and S.K. Jain, eds), pp. 305-329 London: Chapman & Hall.

Linhart, Y.B. and Grant, M.C. (1996) Evolutionary significance of local genetic differentiation in plants. *Annual Review of Ecology and Systematics*, **27**: 237-277.

Lohmeyer, W and Sukopp, H. (1992) Agriophyten in der Vegetation Mitteleuropas. *Schriftenreihe für Vegetationskunde*, Vol. 25, Bonn-Bad Godesberg: Bundesforschungsanstalt für Naturschutz und Landschaftsökologie.

Lynch, M. and Walsh, B. (1998) Genetics and Analysis of Quantitative Traits, Sunderland: Sinauer Associates, Inc. Publishers.

McGraw, J.B. and Antonovics, J. (1983) Experimental ecology of *Dryas octopetala* ecotypes. 1.Ecotypic differentiation and life-cycle stages of selection. *Journal of Ecology,* **71**: 879-897.

Monzingo, H.N. (1987) Shrubs of the Great Basin, Reno: University of Nevada Press.

Parker, I.M., Rodriguez, J. and Loik, M.E. (2003) An evolutionary approach to understanding the biology of invasions: Local adaptation and general-purpose genotypes in the weed *Verbascum thapsus. Conservation Biology,* **17**: 59-72.

Primack, R.B. and Kang, H. (1989) Measuring fitness and natural selection in wild plant populations. *Annual Review of Ecology and Systematics,* **20**: 367-396.

Rajakaruna, N., Siddiqi, M.Y., Whitton, J., Bohm, B.A. and Glass, A.D.M. (2003) Differential responses to Na+/K+ and Ca2+/Mg2+ in two edaphic races of the *Lasthenia californica* (Asteraceae) complex: A case for parallel evolution of physiological traits. *New Phytologist,* **157**: 93-103.

Rice, K.J. and Mack, R.N. (1991) Ecological Genetics of *Bromus tectorum.* 3. the Demography of Reciprocally Sown Populations. *Oecologia,* **88**: 91-101.

Richards, C.L., Bossdorf, O., Muth, N.Z., Gurevitch, J. and Pigliucci, M. (2006) Jack of all trades, master of some? On the role of phenotypic plasticity in plant invasions. *Ecology Letters,* **9**: 981-993.

Sakai, A.K., Allendorf, F.W., Holt, J.S., Lodge, D.M., Molofsky, J., With, K.A., Baughman, S., Cabin, R.J., Cohen, J.E., Ellstrand, N.C., McCauley, D.E., O'Neil, P., Parker, I.M., Thompson, J.N. and Weller, S.G. (2001) The population biology of invasive species. *Annual Review of Ecology and Systematics,* **32**: 305-332.

Sawada, S., Nakajima, Y., Tsukuda, M., Sasaki, K., Hazama, Y., Futatsuya, M. and Watanabe, A. (1994) Ecotypic differentiotion of dry matter production processes in relation to survivorship and reproductive potential in *Plantago asiatica* populations along climatic gradients. *Functional Ecology,* **8**: 400-409.

Schlichting, C.D. (1986) The evolution of phenotypic plasticity in plants. *Annual Review of Ecology and Systematics,* **17**: 667-693.

Schlichting, E., Blume, H.-P. and Stahr, K. (1995) Bodenkundliches Praktikum, Berlin: Blackwell Wissenschafts-Verlag.

Sexton, J.P., McKay, J.K. and Sala, A. (2002) Plasticity and genetic diversity may allow Saltcedar to invade cold climates in North America. *Ecological Applications,* **12**: 1652-1660.

Snaydon, R.W. (1961) Competitive ability of natural populations of *Trifolium repens* and its relation to differential response to soil factors. *Annals of Human Genetics*, **25**: 177.

Sokal, R.R. and Rohlf, F.J. (1995) Biometry, New York: W.H. Freeman and Company.

Stockwell, C.A., Hendry, A.P. and Kinnison, M.T. (2003) Contemporary evolution meets conservation biology. *Trends in Ecology & Evolution*, **18**: 94-101.

Thompson, J.N. (1998) The population biology of coevolution. *Researches on Population Ecology*, **40**: 159-166.

Turkington, R. and Harper, J.L. (1979) Growth, distribution and neighbor relationships of *Trifolium repens* in a permanent pasture. 4. Fine-scale biotic differentiation. *Journal of Ecology*, **67**: 245-254.

van de Laar, H.J. (1975) *Mahonia en Mahoberberis*. *Dendroflora*, **11/12**: 19-33.

Ward, J.K., Antonovics, J., Thomas, R.B. and Strain, B.R. (2000) Is atmospheric CO2 a selective agent on model C-3 annuals? *Oecologia*, **123**: 330-341.

Waser, N.M. and Price, M.V. (1985) Reciprocal transplant experiments with *Delphinium nelsonii* (Ranunculaceae): Evidence for local adaptation. *American Journal of Botany*, **72**: 1726-1732.

Weaver, S.E. and Dirks, V.A. (1984) Variation and climatic adaptation in northern populations of *Datura stramonium*. *Canadian Journal of Botany*, **63**: 1303-1308.

Weber, E. and Schmid, B. (1998) Latitudinal population differentiation in two species of *Solidago* (Asteraceae) introduced into Europe. *American Journal of Botany*, **85**: 1110-1121.

Wieneke, S., Prati, D., Brandl, R., Stöckling, J. and Auge, H. (2004) Genetic variation in *Sanguisorba minor* after 6 years in situ selection under elevated CO2. *Global Change Biology*, **10**: 1389-1401.

Zeitlhöfler, Andreas (2002) *Mahonia aquifolium* - Die Gemeine Mahonie. http://www.garteninfos.de/wildobst/Dipl4-11.html.